SYSTEMS FOR INSTRUCTIONAL IMPROVEMENT

SYSTEMS FOR INSTRUCTIONAL IMPROVEMENT

Creating Coherence from the Classroom to the District Office

Paul Cobb, Kara Jackson,
Erin Henrick, Thomas M. Smith,
and the MIST Team

HARVARD EDUCATION PRESS
Cambridge, Massachusetts

Second Printing, 2021

Paperback ISBN 978-1-68253-177-8
Library Edition ISBN 978-1-68253-178-5

Library of Congress Cataloging-in-Publication Data
Names: Cobb, Paul, 1953- author. | Jackson, Kara, (Co-author of Systems for instructional improvement), author. | Henrick, Erin C., author. | Smith, Thomas M. (Thomas Max), author. | MIST Project.
Title: Systems for instructional improvement : creating coherence from the classroom to the district office / Paul Cobb, Kara Jackson, Erin Henrick, Thomas M. Smith, and the MIST Team.
Description: Cambridge, Massachusetts : Harvard Education Press, 2018. | Includes bibliographical references and index.
Identifiers: LCCN 2017059824| ISBN 9781682531778 (pbk.) | ISBN 9781682531785 (library edition)
Subjects: LCSH: School improvement programs–United States. | Educational change–United States. | Mathematics–Study and teaching–United States. | Educational tests and measurements–United States. | Teachers–Training of–United States. | Teaching–United States–Evaluation.
Classification: LCC LB2822.82 .S96 2018 | DDC 371.2/07–dc23
LC record available at https://lccn.loc.gov/2017059824

Published by Harvard Education Press,
an imprint of the Harvard Education Publishing Group

Harvard Education Press
8 Story Street
Cambridge, MA 02138

Cover Design: Ciano Design
Cover Image: shuoshu/DigitalVision Vectors/Getty Images

The typefaces used in this book are ITC Legacy Serif and ITC Legacy Sans.

Table of Contents

ONE

Investigating and Supporting Instructional Improvement

PAUL COBB, KARA JACKSON, ERIN HENRICK,
AND THOMAS M. SMITH

SCHOOLS AND DISTRICTS across the United States face the challenge of supporting increasingly diverse groups of students to attain increasingly rigorous learning goals across multiple subject matter areas. The *Common Core State Standards* in mathematics and in English language arts and the *Next Generation Science Standards* specify more rigorous goals for students' learning and have influenced state standards and assessments.[1] These new standards and assessments, which began being introduced in 2011, do not merely change the grade levels at which content is taught but also require students to develop new capabilities in those content areas. In the case of mathematics, the previous state assessments were developed in response to *No Child Left Behind* and focused on whether students could solve familiar tasks by performing procedures efficiently and accurately.[2] In contrast, new, more rigorous assessments such as those developed by the Smarter Balanced Assessment Consortium (SBAC) and the Partnership for Assessment of Readiness for College and Careers (PARCC) include tasks that require students to analyze novel problems in order to figure out the procedures they need to perform.[3] Not surprisingly, the proportion of students judged to be proficient dropped significantly in most districts when the more rigorous assessments were implemented.[4]

These changes in standards and assessments have immediate implications for teachers' classroom practices. For example, typical US mathematics instruction emphasizes procedures at the expense of conceptual understanding and problem solving.[5] However, to teach to these standards, teachers will need to develop what the mathematics education community calls *ambitious instructional practices* that support students' development of both conceptual understanding and procedural fluency.[6] As a further challenge, classroom instruction is expected to be equitable and support the learning of all students, including students who have been historically underserved in US schools. Research on teacher learning and teacher professional development makes it clear that developing instructional practices that are both ambitious and equitable is challenging for most US teachers and requires substantial support for an extended period of time.[7]

In the Middle School Mathematics and the Institutional Setting of Teaching (MIST) Project on which this book is based, we partnered with four large urban districts, two districts for four years, and two districts for eight years, to investigate what it takes to support teachers' development of ambitious and equitable instructional practices on a large scale. We speak of *partnering* with the districts and of *collaborating* with district personnel to highlight that we did research *with* rather than *on* the districts and attempted to add value to their instructional improvement efforts. The findings that we report provide actionable, research-based guidance on the types of improvement strategies that schools and districts can implement to address the challenge of supporting all students in attaining increasingly rigorous learning goals. Although our work focused on mathematics in the middle grades, one of our partner district's efforts to apply our findings to other grade levels and to other core subject matter areas suggests that the findings are relevant to instructional improvement efforts that aim at rigorous learning goals more generally.

A SYSTEMIC PERSPECTIVE ON TEACHING AND STUDENT LEARNING

It will become apparent when we share our findings and discuss their implications that there is no silver bullet for improving the quality of instruction at a large scale: no single strategy is, by itself, likely to support instructional improvements that will enable large numbers of students to meet more rigorous learning standards. Our findings indicate that it is essential to take a broad perspective that spans from the classroom to the district central office. For example, a range of improvement strategies for supporting teachers' learning, including pull-out professional development, regular time for collaboration, and content-focused coaching, can each make important contributions. However, the effectiveness of each of these strategies in supporting teachers' learning increases when they are coordinated and become facets of a teacher learning system. Furthermore, the impact of this system of supports is enhanced or impeded by features of the school and district contexts in which teachers work, such as the expectations that school leaders communicate to teachers about what to teach and how to teach it.

We can best illustrate the value of a broad systemic perspective by giving an example that might, at first glance, seem far removed from the world of classroom teaching and student learning. The example is often used as a case study in business schools and concerns a joint venture between General Motors (GM) and Toyota that began in 1984 with the opening of the New United Motor Manufacturing Inc. (NUMMI) car plant in Fremont, California.[8] GM hoped to learn the Toyota production system for building high-quality cars and Toyota hoped to learn how to build cars in the United States, an important consideration as the US Congress had indicated that it might restrict car imports.

Our motivation for sharing this example is *not* to suggest that schools and districts should be modeled on efficient businesses or that their management should be privatized. Teaching that aims to support diverse groups of students in attaining ambitious learning goals is far more complex and demanding work than building cars on an assembly line. While the NUMMI plant was producing high-quality cars reliably after only four months, research on teacher learning indicates that teachers' development of ambitious and equitable instructional practices requires two to three years of sustained support.[9] Nonetheless, even though the success of NUMMI was the result of a relatively straightforward process compared to supporting the improvement of teaching and learning across a school or district, it required a systemic approach. The example is useful in clarifying the scope of the system that might be required to support instructional improvement.

The Toyota production system that was implemented in the NUMMI plant differed significantly from GM's production system by emphasizing quality over quantity, teamwork, and empowering workers. A fundamental principle of the GM system was to maximize output by ensuring that the production line never stopped. Flaws that were identified were only repaired after cars came off the assembly line. In contrast, the Toyota system prioritized quality, and workers were expected to stop the line if they identified a problem. Thus, in the GM system, assessment was a separate event that occurred after production, the equivalent of end-of-unit and end-of-year assessments. In contrast, assessment was integral to and informed improvements in the production process in Toyota system, the equivalent of ongoing formative assessment.

NUMMI workers and managers required support to learn the Toyota system. The key point to note is that the supports for their professional learning were organized into a coherent system that aimed at a single set of work practices. These supports included a two-week visit to a Toyota plant in Japan to work alongside Japanese colleagues, the equivalent of pull-out teacher professional development. Workers were then organized into teams of four or five, the equivalent of teacher collaborative groups, which provided on-the-job support in solving problems encountered and improving the quality of production. Furthermore, each team included a team leader, the equivalent of a coach or teacher leader, who stepped in to provide additional support when problems arose.

The NUMMI case indicates the value of intentionally designing supports for teachers' professional learning as a coherent system from the outset. This insight is especially relevant to instructional improvement given that the professional learning required for instructional improvement is significantly more challenging than that required for the NUMMI workers to learn the Toyota production system. Although our criteria for recruiting our four partner districts included that district leaders had begun to take a systems perspective and were investing in supports for teacher's learning, we found that the supports they implemented were rarely coordinated and often focused on different aspects of instruction.

Given that Toyota had apparently shown GM the secret to its success, it might at first seem surprising that GM was unable to capitalize on what it learned at NUMMI in its other plants. The director of a second GM plant in Van Nuys, California, attempted to reproduce the success of the NUMMI plant by implementing the Toyota production system. In reflecting on the challenges that he encountered, he observed that although he and other GM managers had studied the assembly line at the NUMMI plant, they had failed to consider important conditions that needed to be in place for the assembly line to operate effectively. They had asked questions about Toyota's production system but not about processes and functions beyond the shop floor that supported the line. In other words, Toyota had not in fact shared all the secrets of its success with a competitor.

This is analogous to the leaders of a district focusing only on supports for teachers' professional learning, while overlooking important conditions that shape whether those supports are implemented effectively, including the role of school leaders and district leaders. Our findings indicate that school leaders influence whether teacher collaboration meetings and coaches' work with teachers are productive, and that the lack of coordination between key district central office departments has detrimental impact on instructional improvement efforts.

The value of taking a broad perspective that spans from the classroom to the district central office is that it enables us to attend to the full scope of the system for improving teaching and student learning across a school district, and to how the various parts of this system can be coordinated to work together. Thus, the findings we report bridge research on teaching and learning and research on educational policy and leadership, two areas of investigation that are siloed in most colleges of education. Furthermore, our findings have implications for the district central office departments of Leadership and of Curriculum and Instruction, which frequently pursue conflicting agendas.

It took General Motors twenty-five years and the then largest bankruptcy in US history to figure out how to make higher quality cars reliably. Educators have been trying to figure out how to enable all students to attain worthwhile learning goals for far longer and have made only limited progress. The work that we report focuses directly on this issue and results in the identification of a coherent set of strategies for building school and district capacity for instructional improvement.

OVERVIEW OF THE BOOK

In chapter 2, we describe how we collaborated with two of our partner districts for four years and two for eight years to improve middle grades mathematics instruction. Together, these large urban districts served a total of 360,000 students, with demographics that mirror those of many large urban districts in the United States. As all four districts aimed to support students' attainment of rigorous learning goals in mathematics by improving the quality of instruction, we were able to forge common agendas with them.

These collaborations had both pragmatic and research goals. Our pragmatic goal was to add value to the districts' instructional improvement efforts. Our research goal was to understand what it takes to improve the quality of mathematics teaching and learning on a large scale by identifying productive improvement strategies that would be relevant to other districts aiming at rigorous learning goals for all students.

We addressed our pragmatic goal by collecting and analyzing data each year to document how each district's instructional improvement strategies were actually playing out in schools and classrooms. We then shared our findings with the leaders of each district and made actionable recommendations about how they might revise their current improvement strategies to make them more effective. In chapter 2, we outline the process we followed to provide the districts with timely feedback and recommendations, and describe the framework we developed for analyzing improvement strategies. This framework should be of interest to school and district instructional leaders because it can be used to identify the potential limitations of specific improvement strategies before they are implemented, as well as to explain why the outcomes of implementing particular strategies differed from those intended.

Although our role when making recommendations was advisory, the districts attempted to implement 67 percent our recommendations (see chapter 14). Furthermore, the two districts with which we collaborated for eight years saw a gradual improvement in student's achievement on their state's annual mathematics assessments. This was the case even when state tests increased in rigor.

In the course of our collaboration with the four districts, we compiled a longitudinal data set that spans from the classroom to district instructional leadership. We describe how we selected schools in which to collect data, the number of teachers, coaches, school leaders, and district leaders who participated, and the nature of data we collected. The retrospective analyses that we conducted of these data enabled us to address our primary research question: *what does it take to improve the quality of math teaching and student learning on a large scale?* The resulting findings that we discuss and illustrate in this book provide specific guidance for the development of school and district instructional improvement plans that aim at rigorous learning goals for all students.

In the next set of chapters, we share our findings and make actionable recommendations about potentially productive instructional improvement strategies. The strategies and the findings on which they are based together constitute what Argyris and Schön termed a *Theory of Action* (ToA) for instructional improvement at scale.[10] This ToA consists of:

- A set of specific, actionable improvement strategies
- A research-based rationale that explains why it is reasonable to expect that the strategies will support the development of ambitious and equitable instructional practices

In turn, the ToA that we propose consists of three top-level components: a *coherent instructional system* for supporting teachers' improvement of their instructional practices, *school leaders' practices as instructional leaders* in mathematics, and *district leaders' practices in supporting the development of school-level capacity* for instructional improvement.[11]

As shown in figure 1.1, the first of these components, a coherent instructional system, comprises four elements:

- Goals for students' mathematics learning and vision of high-quality mathematics instruction
- A subsystem of supports for teachers' learning that includes pull-out professional development, school-based teacher collaborative meetings, and mathematics coaching, as well as teachers' advice networks
- Instructional materials and assessments
- Supplemental supports for currently struggling students

Chapters 3–11 focus on the various elements of the coherent instructional system. In each chapter, we clarify why the element in question matters for instructional improvement at scale and give an overview of the challenges that our partner districts encountered. We then discuss the findings from our own and others' work that provide a research-based rationale for the actionable implications or takeaways for school and district instructional leaders. Finally, each chapter briefly outlines next steps for research that aims to inform school and district instructional improvement efforts.

FIGURE 1.1 The coherent instructional system

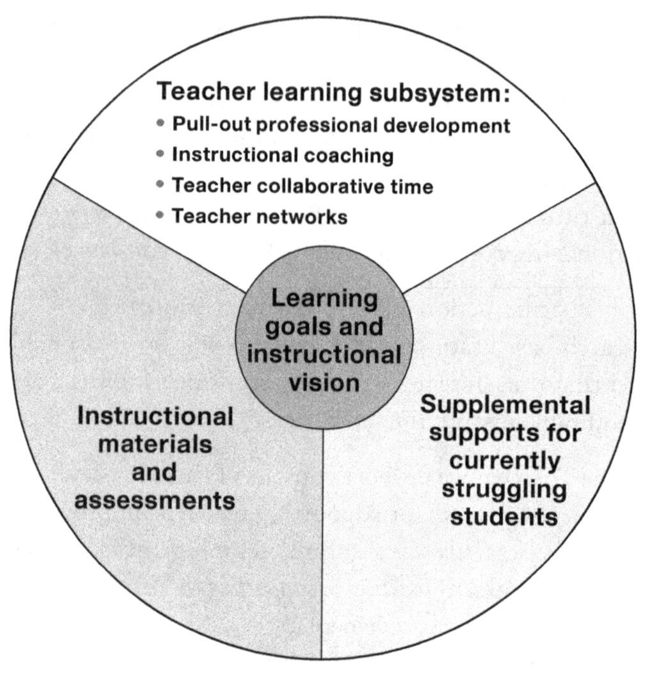

As shown in figure 1.1, *goals for students' learning* and a corresponding *vision of what counts as high-quality mathematics instruction* are at the center of the coherent instructional system. In chapter 3, we first clarify what is meant by rigorous learning goals in mathematics. We then draw on current research on teaching and learning to specify a vision of ambitious and equitable instruction that enables all students, including those from historically disadvantaged groups, to attain those goals. This vision of high-quality instruction implies specific goals for teachers' improvement of their instructional practices, which can be translated into goals for teachers' learning. It is therefore essential that the remaining elements of the coherent instructional system shown in figure 1.1 are aligned with the core element of learning goals and the associated instructional vision if they are to support teachers' development of instructional practices that enable *all* students to attain rigorous learning goals.

In chapter 3, we also share findings that indicate that although deep content knowledge for teaching is important, it is not sufficient if teachers are to develop ambitious and equitable instructional practices. These findings provide strong evidence that teachers also need to develop a relatively sophisticated image or vision of what high-quality instruction looks like, and to realize that all their students can engage in rigorous mathematical activity if they are given appropriate support. These findings are important with regard to equity and serve to further specify goals for teachers' learning. For most teachers in our four partner districts, and for most teachers in the United States more generally, the learning involved is substantial and requires support that is both extensive and sustained.

Chapter 4 introduces a key element of a coherent instructional system, the *teacher learning subsystem*, comprised of four types of support for teachers' development of ambitious and equitable instructional practices. Our partner districts all implemented pull-out professional development that was organized at the district level, school-level teacher collaborative meetings, and instructional coaching. The findings that we report indicate that it is crucial to connect these types of supports by ensuring that they have a common focus and aim to support the improvement of the same specific aspects of instruction. The teacher learning subsystem also includes a fourth type of support, teachers' advice networks, that is not usually considered when thinking about instructional improvement. However, our findings indicate that the advice that teachers receive from colleagues can indeed support them in improving the quality of their instruction. In the chapter, we illustrate what a potentially productive teacher learning subsystem looks like and propose principles that can guide the design of a coordinated system of supports for teachers' learning.

The next four chapters each focus on one of the four parts of the teacher learning subsystem shown in figure 1.1. As we clarify in chapter 5, *pull-out professional development* refers to professional development in which teachers from multiple schools engage in professional learning. We discuss the potential of pull-out professional development but also summarize findings of prior research that indicate that high-quality pull-out professional development is, by itself, often insufficient to support teachers' development of ambitious and equitable instructional practices. These findings further emphasize the importance of coordinating supports for teachers' learning. We therefore treat pull-out professional development as one facet of a teacher learning subsystem when we make recommendations for its design and facilitation.

The second facet of the teacher learning subsystem, *regularly scheduled time for teachers to collaborate*, has the potential to support teachers in deepening their content knowledge, developing productive views of their students' current mathematical capabilities, and improving their classroom practice. However, as we discuss in chapter 6, the potential of teacher collaborative meetings is frequently not realized. We describe key differences between more and less effective teacher collaboration, illustrate what effective collaboration looks like, and clarify how school leaders can influence teacher collaboration. In doing so, we highlight that in effective collaborations, teachers investigate the types of reasoning their students are developing and relate what the students are learning to the instruction that they received. We then discuss a number of different challenges to designing and implementing effective teacher collaboration and make recommendations for addressing these challenges.

Instructional coaching, as we define it in chapter 7, involves teachers working with a person with content-specific pedagogical expertise who is charged with supporting them to improve the quality of their instruction. A positive feature of this definition is that it encompasses the work of teacher leaders as well as that of formally designated coaches. The promise of instructional coaching as a primary support for teachers' learning stems from its potential to ground teachers' development of ambitious and equitable instructional practices in their daily work with students. Our partner districts implemented a range of different content-specific coaching programs, thus giving us the opportunity to investigate the strengths and weaknesses of each in realizing this promise.

In discussing our findings, we first focus on the types of coaching activities that have the potential to support teachers' development of ambitious and equitable instructional practices. We then clarify what coaches need to know and be able to do beyond being accomplished teachers to implement these activities effectively with teachers. The various aspects of coaching expertise that we discuss constitute goals for coaches' learning and can thus orient the design of professional development for coaches. Finally, we discuss features of school and district contexts that influence whether coaches actually spend the bulk of their time working with teachers on instructional issues, and draw attention to both the critical role of the principal and the trade-offs between school-based and district-based coaching models.

In chapter 8, we focus on the final facet of the teacher learning subsystem, *teachers' advice networks*. Advice networks differ from the other three types of supports in that school and district leaders cannot directly design them or mandate them into being. Instead, they emerge as individual teachers decide to turn to a colleague for advice about instruction. We first share our findings about the influence of teachers' interactions with colleagues, highlighting that interactions with colleagues who have developed more ambitious instructional practices can support improvements in teachers' own instructional practices. This finding serves to underscore the significant but often unnoticed contribution of teachers' networks to instructional improvement. We next report our findings about how other elements of the theory of action including coaching, teacher collaborative time, and school instructional leadership can influence teachers' decisions about whether and to whom to turn to for advice about instruction. These findings clarify that although advice networks cannot be directly designed, school and district leaders can influence the development of productive networks indirectly by ensuring that pull-out professional development, teacher collaborative time, and content-focused coaching are of high quality, and by improving their own leadership practices. In doing so, they foster an important source of support for teachers' learning that they, in effect, get for free.

How the implementation of *instructional materials*, such as a textbook series and district curriculum frameworks, contributes to the coherent instructional system is the focus of chapter 9. We define high-quality instructional materials as those designed to support students in achieving rigorous learning goals by providing challenging tasks that are sequenced to support students' development of significant mathematical ideas. Each of our four partner districts adopted high-quality materials, although there was variability in both the extent to which teachers used the materials and the degree to which they implemented them effectively. Our analyses point to the importance of district-wide or at least school-wide adoption of high-quality materials as well as professional development that is organized around the materials.

In chapter 10, we focus on the critical role of *state and district assessments* in the coherent instructional system. Although state assessments had the potential to provide teachers with feedback on their students' learning, their lack of alignment with the learning goals of our partner districts contributed to a tension in district and school leaders' goals and practices. Our findings on how teachers and leaders made use of assessment data throughout the MIST study highlight the challenge of supporting the implementation of ambitious and equitable instruction when state assessment systems privilege less rigorous learning goals.

Chapter 11 focuses on the final element of a coherent instructional system, the provision of *supplemental supports for students who are currently struggling* in mathematics. We limit our analysis to an extremely common type of supplemental support that schools in all four partner districts provided, so-called "double dose" or second math classes. These classes typically emphasized procedural skills rather than conceptual understanding and problem solving in an attempt to increase student achievement on end-of-year state assessments. Further, we found that the supplemental classes were rarely effective even though the state assessments at the time focused primarily on procedural capabilities. Our findings lead us to propose that both the second and the mainstream classes should aim at rigorous learning goals and that the immediate goal of supplemental supports should be to enable students to participate fully and learn in the context of their mainstreams classes. This, in turn, implies that the teachers of the two classes should meet regularly to plan and coordinate their instruction.

David Cohen, a leading researcher of educational policy, indicated the relevance of the above chapters to schools and districts aiming at rigorous learning goals for all students when he noted that "schools and districts across the country are trying to develop coherent instructional systems but they don't know what one actually looks like."[12] Taken together, the findings and recommendations that we discuss in these chapters give a relatively detailed image of a coherent instructional system that has at its center rigorous learning goals for all students.

The next two chapters focus on the remaining two components of the ToA: school instructional leadership and district instructional leadership. As we make clear in chapter 12, a primary goal for school instructional leadership should be to initiate and guide the development of a coherent instructional system, thereby creating conditions for teachers' ongoing improvement of their instructional practices. Although observing classroom instruction and providing feedback that communicates appropriate instructional expectations is a common aspect of instructional leadership in many districts, our analysis of school leaders in the MIST districts suggests that this is extremely challenging, even for school leaders who have received considerable professional development. We also found that school leaders who collaborated with a district math specialist or coach were more successful in creating sustained opportunities for teachers to improve the quality of their instruction. Based on these findings, we propose that districts support principals in understanding the connections between student learning goals, teachers' instructional practices, and teacher learning, and that districts foster collaborations between school leaders and colleagues with expertise in ambitious and equitable instruction.

In chapter 13, we first describe the tendency of district leaders in the offices of Leadership and of Curriculum and Instruction to pursue incompatible agendas. Principal Supervisors typically emphasized improving student achievement on procedurally oriented state assessments in the short term whereas leaders in Curriculum and Instruction typically emphasized supporting teachers' development of ambitious and equitable instructional practices in the long term. Crucially, we found that this conflict in agendas adversely affected our partner districts' instructional improvement efforts. Against this background, we share findings that clarify what might be necessary for the two groups of leaders to establish a shared agenda organized around rigorous learning goals and an ambitious and equitable vision of instruction. In doing so, we propose that a primary goal of district instructional leadership should be to support the development of school-level capacity for instructional improvement.

There is much talk of transforming schools and districts into learning organizations. However, it makes little sense to talk of organizational learning unless we can distinguish between mere change and improvement.[13] Change is endemic to schools and districts, but most of these changes are not improvements.[14] The theory of action outlined above specifies a primary goal or target for school and district organizational learning, and can therefore orient and inform their efforts not merely to change but to improve.

In the final two chapters, we assess the impact of our annual feedback and recommendations to districts and pull out themes that cut across the various components of the theory of action. Chapter 14 reports an analysis of the extent to which district leaders acted on our recommendations and whether they were able to implement the recommendations successfully. Whereas our partner districts took up about two-thirds of our recommendations, they frequently struggled to implement them even though the recommendations focused on types of improvement strategies that are being widely implemented. We also examine the factors that influenced both the districts' uptake of our recommendations and the extent to which they implemented them successfully.

In chapter 15, we examine themes that cut across the various chapters. These include the value of taking a systems perspective that involves formulating improvement strategies by mapping out from the classroom, the importance of identifying and capitalizing on the capabilities of school and district personnel with expertise in ambitious and equitable instruction and in supporting teachers' learning, and the pervasive influence of state accountability assessments. We then discuss the implications of our work for developing instructional improvement plans that are customized to particular schools or districts and also consider what might be involved in adapting the theory of action to other content areas. Finally, we focus on the implications of our work for research that can inform instructional improvement. In doing so, we suggest that an important next step after identifying potentially productive improvement strategies is to investigate what it takes to implement the strategies reliably in a range of different school and district contexts. In addition, we consider what it might take to bridge the current schisms between research on classroom teaching and learning and on educational policy and leadership, as well as the more general schism between research and practice.

While the prior chapters reflect the research team's perspective on our district partnerships, the Afterword provides a practitioner's account of our eight-year partnership with one of our partner districts. Here, Michael Sorum describes some of the MIST findings and recommendations that proved to be particularly salient for his district and illustrates how they influenced the revision of the district's improvement strategies. He also discusses what he finds valuable in Research Practitioner Partnerships and what it takes to make them productive.

We have also included an appendix in which two external researchers, Annie Allen and William Penuel, report findings from their study of our partnerships with two of the districts for two years. In doing so, they clarify both how district leaders viewed their partnership with MIST researchers and how the leaders assessed the value of the partnership work for their districts. Their findings clarify the extent to which the leaders in each district had come to trust MIST researchers and to view us as concerned with issues facing their districts.

TWO

Investigating Instructional Improvement in Partnership with Districts

PAUL COBB, ERIN HENRICK, KARA JACKSON, AND THOMAS M. SMITH

IN THIS CHAPTER, we describe how we partnered with the four districts both to add value to their instructional improvement efforts and to develop a theory of action for instructional improvement at scale that is relevant to other schools and districts attempting to support all students' attainment of rigorous learning goals. We first outline the research base on which we could draw when working with the four partner districts and then give background information on the districts. Next, we describe the process we followed each year to provide the four districts with timely feedback about how their instructional improvement strategies were actually being implemented in schools and classrooms. In doing so, we share the framework we used to develop actionable recommendations about how district leaders could revise their instructional improvement strategies to make them more effective. Finally, we describe the approach we took to achieve our research goal of developing a theory of action for instructional improvement at scale.

THE RESEARCH BASE

Over the past twenty-five years, research in mathematics education and related fields has made considerable progress in documenting long-term student learning trajectories in different mathematical domains (e.g., algebra, statistics) that culminate with their development of a relatively deep understanding of central mathematical ideas.[1] Additionally, instructional materials have been published that build on this research on student learning.[2] Recent research has also delineated specific instructional practices that support students' attainment of rigorous learning goals.[3] In addition, there has been considerable progress in understanding professional development that supports teachers' development of practices aimed at rigorous learning goals for students.[4]

Although these advances are significant, they have had limited influence on instruction in most classrooms, which continues to focus primarily on performing procedures at the expense of conceptual understanding, reasoning, and problem solving.[5] In our view, this limited impact is due in large measure to the way in which research on teaching and learning typically treats classrooms and teachers as disconnected from school and district policies and leadership. Yet, abundant evidence suggests that teachers' instructional practices are profoundly influenced by the school and district contexts in which they work.[6]

It might seem reasonable to turn to research on educational policy and leadership for guidance on improving the quality of instruction and student learning on a large scale. However, research in these areas can also provide only limited assistance. This is in part because this research does not typically take a position on the types of student learning goals that are worth pursuing.[7] The findings that we share in later chapters indicate that a focus on rigorous learning goals and thus on teachers' development of ambitious and equitable instructional practices has implications for effective school and district instructional leadership (see chapters 12 and 13). In addition, much of the research on educational policy and leadership assesses whether either the implementation of particular policies or leaders' adoption of particular practices (the inputs) impacts student outcomes (the outputs). In these studies, the classroom is treated as a black box whose internal processes, classroom teaching and learning, are not examined.

From a systems perspective, the disconnect between research on teaching and learning and research on policy and leadership is extremely problematic.

There were major gaps in the research base on which we could build when we began collaborating with the four districts, and we found that relevant research became increasingly thin as we moved out from the classroom.[8] Not surprisingly, we also found that school and district personnel often had to make decisions on which research could provide little guidance as they strove to support teachers' development of ambitious and equitable instructional practices. These weaknesses in the research on which we could draw underscore the need to frame the challenges inherent in supporting instructional improvement at scale as a researchable issue. We contend that this issue is best investigated by partnering with practitioners to conduct research with rather than on them.

DISTRICT PARTNERSHIPS

Research-Practice Partnerships (RPPs) in which researchers and practitioners engage in long-term collaborations to address pressing problems of practice have become increasingly common in recent years.[9] However, at the time that we partnered with the four districts, the number of RPPs was small, and most concentrated on evaluating district policies and programs to determine if they were effective, thereby complementing the work of district research and evaluation departments.[10] MIST was an atypical RPP in that it focused directly on classroom teaching and learning in a particular content area. In doing so, MIST researchers collaborated with district leaders across five central office departments to design, study, and revise instructional improvement strategies organized around rigorous learning goals and an ambitious and equitable vision of instruction (see chapter 3).[11]

Partner Districts

As indicated in chapter 1, we collaborated with four urban districts for four years (2007–2011) and then extended our collaboration with two of the districts for an additional four years (2011–2015). The original four districts served a total of 360,000 students, and the two districts that we worked with for an additional four years served a total of 180,000 students. We extended our collaborations with two of the districts so that we could work more closely with school and district leaders to co-design and co-lead pull-out professional development for principals and for instructional coaches who were expected to work directly with teachers (see chapter 5). All four partnerships focused on improving the quality of instruction and student learning in middle-grades mathematics.

The four districts were typical urban districts in the United States in many respects. For example, they had limited financial resources, served a relatively high proportion of students from traditionally underserved groups, were attempting to reduce large achievement and opportunity gaps, and had identified improving student achievement in middle-grades mathematics as a priority. In addition, three of the four districts had high teacher turn over and a relatively large proportion of novice teachers.

However, the districts were atypical in several important respects. It was essential that we forge common improvement agendas with the districts and thus establish genuine partnerships that were grounded in trust.[12] We therefore recruited districts that were responding to the high-stakes accountability demands of the No Child Left Behind Act in 2001 (NCLB) by attempting to improve the quality of instruction rather than by teaching to procedurally-oriented state assessments.[13] All four districts were attempting to support all students' attainment of rigorous learning goals and understood that this would require significant improvements in the quality of classroom instruction. They were therefore providing professional development and other supports for teachers' development of ambitious instructional practices. Furthermore, three of the four districts had adopted commercially published instructional materials for middle-grades mathematics that aimed at rigorous learning goals, and the fourth was supplementing the traditional materials it had adopted by writing lessons organized around tasks that were of high cognitive demand. In addition, all four districts were attempting to support principals' development as instructional leaders in mathematics and in other core content areas.

To maintain the confidentiality of the districts, we use pseudonyms (A–D) that reflect the order in which we recruited them. Table 2.1 provides information for each of the districts regarding the length of our partnership; the region of the country in which the district was located; the number of students in year one of the MIST project (2007–2008); and student demographics provided by the districts for year one of the project, including racial/ethnic background, number of students identified as English learners, and percentage of students eligible for free or reduced-price lunch. In what follows, we provide additional information about the districts, including their initial improvement strategies and how those evolved, changes in

TABLE 2.1 Characteristics of partner districts, 2007–2008

	A		B		C		D	
Length of partnership	4 years		8 years		4 years		8 years	
U.S. region	Midwest		Southwest		Southwest		Southeast	
Number of students (rounded to nearest 5,000)	35,000		80,000		160,000		100,000	
Students' racial/ethnic identification (rounded to the nearest percent, so percents may not equal 100%)	African American	40%	African American	26%	African American	29%	African American	35%
	Hispanic	17%	Hispanic	58%	Hispanic	65%	Hispanic	5%
	White	30%	White	14%	White	5%	White	55%
	Asian	9%	Asian	2%	Asian	1%	Asian	2%
	Native American	5%	Native American	0%	Native American	0%	Native American	0%
% students identified limited English proficient	25%		28%		33%		5%	
% students eligible for free or reduced-price lunch	65%		70%		85%		58%	

districts' leadership and in state assessments over the course of the project, and the experience level of the teachers who participated in the project.

District A

District A had the longest history of using reform-oriented mathematics textbooks, as compared to our other three partner districts. The district adopted the inquiry-oriented text, *Connected Mathematics Project 2* (CMP2), in year two of the MIST project.[14] However, district teachers had been using the first edition of *Connected Mathematics Project* for approximately ten years prior to the adoption of CMP2. As described in chapter 9, CMP2 is a problem-centered curriculum designed to support in-depth investigations that are organized around sequenced sets of challenging mathematical tasks. It aims to support students to develop conceptual understanding of key mathematical ideas, problem-solving capabilities, and procedural fluency.

Central office leadership in District A was stable during the four years of our partnership, with the exception that the superintendent retired mid-way through the partnership and the then-current chief academic officer was promoted to superintendent. The state assessment, which was procedurally oriented, was consistent throughout our four years working together.

At the start of the partnership, District A had the most experienced teachers. Of the thirty teachers we randomly selected to participate in the project, only about nine percent had taught for less than three years. District A teachers were, on average, the most accomplished, as compared to teachers in other districts in terms of their instructional practices and mathematical knowledge for teaching.

Across the four years of the partnership, teachers were provided with substantial pull-out professional development that aimed to support their development of a vision of ambitious and equitable mathematics instruction. For example, professional development focused on organizing instruction around challenging tasks and on improving the quality of discourse in the classroom. District A implemented teacher collaborative time starting in year two of the partnership, but it varied in the extent to which it was specific to mathematics (e.g., in some schools, teachers met by grade level not by content area). Mathematics-specific coaching was not a primary instructional improvement strategy in District A schools. In year three of the partnership, the district funded the new position of "instructional facilitator" in each school that was not content-specific. Instructional facilitators were expected support principals as instructional leaders and tended to assist principals in planning, leading, and assessing school-based professional learning.

As was the case in all four districts, central office leaders in District A emphasized the role of principal as instructional leader. Principals were expected to observe classroom instruction for two hours a day and provide teachers with feedback that communicated appropriate instructional expectations. In addition, principals were also expected to assess the overall quality of instruction in their schools by observing multiple classrooms regularly, joined by teachers and occasionally by district leaders. In years one and two of the partnership, principals received professional development on high-quality instruction that was not content specific. In year three, principals participated in one mathematics-specific professional development session (in addition to other non-content specific professional development). In year four, a series of mathematics-specific professional development was offered to assistant principals.

District B

District B adopted CMP2 in year one of the partnership. Although there were three changes in the superintend position in District B across the eight years of our partnership, the district's improvement initiatives remained consistent. This consistency was due, in part, to stable leadership in the position of chief academic officer. The rigor of the state mathematics assessment increased in year five of our eight-year partnership with the district (2011–2012).

At the start of the partnership, District B had only a few middle-grades mathematics teachers who were already accomplished in teaching towards rigorous student learning goals, and also had a large number of novice teachers. About one-third of our participants at the start of the partnership had taught for less than three years, and just over half had taught for less than five years.

In the first four years of the partnership, teachers attended CMP2 professional development led by the curriculum developers. In addition, district mathematics specialists provided regular pull-out professional development in all eight years of the partnership. Mathematics-specific teacher collaborative meetings varied by school in the early years of the partnership, but by year five, nearly all mathematics teachers participated in teacher collaborative meetings on a somewhat regular basis. In addition, district mathematics specialists created and revised elaborate curriculum frameworks across the eight years of the partnership.

District B invested heavily in coaching. For the first four years of the partnership, there was a mathematics coach at each middle school who coached for half the day and taught the other half of the day. As we describe in chapter 7, the school-based coaches were not necessarily more accomplished in their instruction than the teachers they were coaching, and it was apparent that the district's investment in this position was not paying off.

Therefore, starting in year five of the partnership, the central office eliminated the school-based coaching position and instead hired a small number district-wide coaches who had substantial expertise both in ambitious and equitable instruction and in supporting teachers' learning. These coaches served all of the middle-grades schools.

Across the eight years, principals (and in years six through eight, assistant principals) were expected to serve as instructional leaders, which included observing instruction regularly and providing feedback to teachers. In addition, they were expected to work closely with coaches to determine teachers' professional learning needs. In the early years of the partnership, principals were also expected to observe a focal aspect of instruction across multiple classrooms regularly, joined by teachers, coaches, and occasionally by district leaders. Similar to Districts A and C, principals participated in content-general professional development the first three years of the partnership. The MIST team partnered with the mathematics department to provide content-specific professional development in the later years of the partnership.

District C

District C had adopted a conventional textbook series. However, the district supplied CPM2 to schools as a secondary resource, and district mathematics specialists wrote extensive curriculum frameworks in which they provided supplemental lessons organized around challenging tasks. Central office leadership remained relatively stable in District C during the four years of the partnership, with one change in the position of superintendent over the four years. District C had a large number of new teachers. About 18 percent of our participants at the start of the partnership had taught for less than three years, and 31 percent had taught for less than five years.

Teacher collaboration was a leading instructional improvement strategy in District C. Across the four years, principals were required to schedule time each school day for mathematics teachers to work together. The designated teacher instructional leaders on each campus was expected to support the principal and lead teacher collaborative time. The district provided pull-out professional development for teachers starting in year two of the partnership, but it was mostly voluntary. Starting in year two of the partnership, the district created the position of district-based coaches; each coach served a number of campuses and was tasked with working one-on-one with teachers in their classrooms. Some coaches also facilitated teacher collaborative meetings, but on many campuses, teacher instructional leaders (who were often less expert than the coaches) continued to serve as facilitators.

District C's expectations for principals as instructional leaders remained the same across all four years of the partnership. They were required to observe classroom instruction frequently and provide feedback to teachers. Similar to Districts A and B, principals were also expected to assess the overall quality of instruction by observing multiple classrooms regularly, joined by teacher instructional leaders and occasionally by district leaders. Principals received consistent and ongoing professional development across the four years of the project, typically in monthly principal meetings. While some of the professional development focused on content-neutral aspects of high-quality instruction (e.g., formative assessments and using academic language), principals also received math-specific professional development each year, ranging from once a year to quarterly.

District D

In contrast to the other three districts, District D's instructional improvement strategies changed significantly during our partnership, which lasted eight years. This was in large part due to major changes in district leadership. District D adopted CMP2 in the first year of the partnership and invested heavily to uspport its implementation. However, there was a change of superintendent and of chief academic officer in year five, and starting in year six, central office leaders no longer officially supported the implementation of CMP2. As described in chapter 9, teacher interviews and surveys as well as video-recorded lessons indicated that as of year six, many teachers used materials they had found on the internet or in sources other than CMP2. In addition, the rigor of the state mathematics assessment increased in year five (2011–2012) just as the average quality of instruction across the district began to decline. Similar to Districts B and C, District D had a large number of novice teachers. About 35 percent of our participants at the start of the partnership had taught for less than three years, and about 54 percent had taught for less than five years.

District D teachers received substantial professional development in the first four years of the partnership, including district-wide professional development that focused on using CMP2 and had opportunities to attend professional development on CMP2 provided by the curriculum developers. Starting in year six, teachers in some schools received mathematics-focused professional development that focused on implementing rigorous tasks and formative assessment practices.

Regularly scheduled time for teacher collaborative meetings was uncommon in the early years of the partnership. However, in year five, it became the leading improvement strategy (however, the frequency of collaborative meetings varied by campus). In addition, a cadre of accomplished district-based mathematics coaches worked with mathematics teachers on campuses for the first five years of the partnership. In year six, the position was discontinued and each school was instead provided with a coach who worked with teachers across all content areas.

Throughout the first five years of our partnership, principals were expected to observe instruction and give teachers feedback, as well as monitor the overall implementation of the district curriculum. However, starting in year six, principals were no longer held accountable for performing these activities. Instead, principals were expected to determine the professional development needs of their staff, and to ensure that teachers met regularly in teacher collaborative time. Across the years of our partnership with District D, principals received monthly content-neutral professional development that focused on their work as instructional leaders. In years two, five, and eight of the partnership, MIST researchers partnered with the mathematics department to offer a few mathematics-specific sessions for principals focused on deepening their vision of high-quality mathematics instruction, and how to observe and provide feedback that communicated appropriate instructional expectations to teachers.

Study Participants

At the start of our partnerships with the four districts in 2007, we worked with leaders of each district to select six to ten middle-grades schools in which we would collect data.[15] Our goal was to focus on schools that were representative of district middle-grades schools in terms of their capacity for improving the quality of instruction. The feedback that we would give district leaders about how their instructional improvement strategies were being implemented would then not simply be stories about a small number of schools, but would have implications for how each district's instructional improvement initiatives were being implemented across all middle-grades schools.

Once we had identified schools, we recruited thirty randomly selected middle-grades mathematics teachers in each district. When a teacher left a particular school, we recruited a replacement teacher from the same school as our focus was on the instructional capacity of the school. In addition, we recruited the principals of these schools and any other school leaders who monitored mathematics instruction, the mathematics coaches who served these schools, and the district leaders from five central office units that had an interest in mathematics instruction and students' learning. These units were leadership, curriculum and instruction, special education, English language learners, and research and evaluation. There were approximately fifty participants in each of the four districts each year, with a total of two hundred participants.

When we reduced the number of partner districts from four to two, we doubled the number of schools and teachers in each district in order to have adequate statistical power for our quantitative analyses. There were then approximately one hundred participants in each district.

LEVELS OF ANALYSES

Our collaboration with the four districts involved conducting data collection, analysis, and feedback cycles in each district each year. Each cycle (discussed in detail below) involved documenting the districts' improvement strategies, collecting and analyzing data to assess how these strategies were being implemented, and reporting the findings to district leaders and making recommendations about how they might revise their strategies. In conducting this work, we differentiated between two levels of analysis that correspond to our pragmatic and research goals. The purpose of the first level of analysis was to add value to the four districts' instructional improvement efforts by providing them with timely evidence of how their strategies were playing out in schools. The purpose of the second level was to develop an empirically grounded theory of action for instructional improvement that is generalizable and could inform instructional improvement in other districts.

Instructional Improvement Strategies as Designs for Supporting Learning

It is sometimes tempting to assume that the failure of an instructional improvement strategy is the result of either willful distortion or resistance, and that increasing incentives and penalties will be sufficient to remedy the situation.[16] However, as David Cohen and Carol Barnes observed, this assumption overlooks the fact that instructional improvement strategies call for implementers to develop new capabilities and thus involve learning.[17] This in turn implies that improvement strategies should include the provision of supports for that learning.[18]

We built on this insight when assessing our partner districts' improvement strategies by distinguishing between what we call the *what* and the *how* of a strategy. The what of a strategy corresponds to the envisioned types of practice that the strategy is intended to support. The how of a strategy refers to both the supports included in the strategy and the rationale for why these supports might be effective. The key question when assessing a strategy before it is implemented and when accounting for how it is being implemented is whether the supports are adequate given the nature of the learning that the strategy calls for. To make this question tractable, we developed a framework that distinguishes between different types of supports and clarifies their potential to scaffold practitioners in improving their practices. The four types of supports that we identified capture all the improvement strategies that the four districts attempted to implement during our partnerships with them: new positions, learning events (including professional development), organizational routines, and tools. We assessed the potential of each type of support by drawing on research on professional learning.[19]

New positions

District instructional improvement strategies often include the creation of new positions whose responsibilities include supporting others' learning by providing expert guidance.[20] For example, one of the districts created the position of a school-based mathematics coach in each middle-grades school whose responsibilities included providing support both to teachers and to principals in their role as instructional leaders in mathematics.

Learning events

Professional development sessions for teachers, coaches, or school leaders are examples of learning events, which we define as scheduled meetings or interactions that can give rise to opportunities for participants to improve their practices. Several distinctions proved useful when assessing our four partner districts' improvement strategies. First, *ongoing intentional learning events* are designed as a series of meetings or interactions that build on each other, and typically involve a relatively small number of participants. Because a small number of participants is involved, the group has the potential to evolve into a genuine community that works together for the explicit purpose of improving their practices.[21] In contrast, *discrete intentional learning events* include one-off professional development sessions as well as a series of learning events that do not build on each other. Although, discrete intentional learning events are, by themselves, unlikely to support major improvements of practice, they have the potential to support the elaboration or extension of current practices (e.g., a single professional development session to familiarize teachers with instructional materials that are consistent with their current classroom practices).[22]

Learning opportunities can also arise *incidentally* when school and district personnel interact with colleagues with greater expertise. For example, teachers often turn to a colleague for advice about instruction. These interactions might support their learning if the advice concerns a significant problem of practice and if the teacher who provides advice is more accomplished.

New organizational routines

Instructional improvement strategies sometimes include the introduction of new organizational routines. Martha Feldman and Brian Pentland define organizational routines as "repetitive, recognizable patterns of interdependent actions, carried out by multiple actors."[23] For example, leaders in one of our partner districts expected that middle school principals would conduct Learning Walks with the mathematics coach at their schools on a regular basis. A Learning Walk is a repetitive, recognizable pattern of actions that involves determining the focus of classroom observations (e.g., the extent to which teachers maintain the cognitive challenge of tasks throughout the lesson), selecting classrooms to visit, observing a classroom, and then conferring to discuss observations before moving on to the next classroom.

Carefully designed organizational routines in which a more expert colleague scaffolds others' learning as they engage in activities that are close to practice can be an important means of supporting learning.[24]

New tools

Instructional improvement efforts almost invariably involve the introduction of a range of new tools. Examples include new instructional materials, curriculum frameworks, classroom observation protocols, and data management systems. Although the use of a new tool can sometimes lead to significant changes in practice, there is also evidence that practitioners often assimilate new tools to their current instructional practices rather than reorganize those practices.[25] It is therefore important to support the introduction of tools whose effective use requires significant changes in practice by designing ongoing intentional learning events.

Summary

When assessing district improvement plans, we anticipated that strategies with the potential to support consequential professional learning would involve some combination of new positions to provide expert guidance, ongoing intentional learning events, carefully designed organizational routines carried out with a more knowledgeable other, and the use of new tools whose effective use is supported. We also took account of the support that discrete intentional learning events and incidental learning events might provide when analyzing the districts' improvement strategies. Because the four types of supports are not specific to our partner districts, the framework can be used to assess the degree to which other schools' and districts' improvement strategies are likely to support professional learning. As we discuss in chapter 15, this framework may therefore be of value to school and district leaders who are attempting to support all students' attainment of rigorous learning goals and thus teachers' development of ambitious and equitable instructional practices.

Pragmatic Goal: Adding Value to Partner Districts' Instructional Improvement Efforts

The data collection, analysis, and feedback cycle that we conducted in each district each year is shown in table 2.2. We discuss each phase of the cycle in turn.

TABLE 2.2 Steps in the annual cycle of data collection, analysis, and feedback cycle

TIMELINE	ACTIVITY
October	Interview key district leaders to document strategies for instructional improvement
October–December	Analyze interviews and create *District Design Document* Share draft *Document* with key district leaders to check for accuracy
January	Interview teachers, coaches, instructional leaders, and district leaders to document the implementation of the strategies
February–April	Analyze interviews Create *District Feedback and Recommendations Report*
May	Share *Report* with district leaders Meet with district leaders to discuss *Report*

Documenting the districts' instructional improvement strategies

Each October, we documented each district's instructional improvement strategies by conducting audio-recorded interviews with approximately ten leaders from each district and also collected relevant documents (e.g., district organizational charts, schedules of professional development sessions, curriculum frameworks, school leader and teacher evaluation forms). In addition to the superintendent and/or chief academic officer, the leaders we interviewed were mainly from the units of curriculum and instruction responsible for supporting mathematics teachers and mathematics coaches, and from the leadership department responsible for supporting and monitoring middle-school leaders. The interviews focused on the district's current goals for middle-grades mathematics instruction and the strategies the district was implementing to achieve these goals. We analyzed transcriptions of these interviews and reported the results in *District Design Documents* of three to four single-spaced pages in which we described each strategy. We shared the *District Design Documents* with district leaders to check whether we had accurately represented their improvement goals and intended strategies, and made revisions if necessary.

Documenting how the districts' instructional improvement strategies were being implemented

We collected multiple types of data in each district each January–March. However, it was essential that we provide district leaders with feedback about how their improvement strategies were being implemented before they began planning strategies for the following school year over the summer. We therefore limited the data we analyzed to assess how the districts' strategies were being implemented to audio-recorded interviews conducted with all two hundred participants in January. The interviews focused on a range of issues including both the formal and informal supports for members of each role group, to whom and for what they perceived themselves to be accountable, and the tools they used in their work (e.g., instructional materials, protocols for teacher collaborative meetings, classroom observations forms). The interview protocols we used for teachers, mathematics coaches, school leaders, and district leaders in different central office units are downloadable at http://vanderbi.lt/mist.[26]

The additional types of data we collected in each district in January–March of each year included:

- On-line surveys for teachers, coaches, and school leaders that focused on similar issues to the interviews. The surveys can be downloaded at http://vanderbi.lt/mist.
- Video-recordings of two consecutive lessons in the 120 participating teachers' classrooms, coded using the Instructional Quality Assessment (IQA).[27] We used eight rubrics to assess the level of challenge or rigor of the tasks used in lessons, the extent to which that level of rigor was maintained in various phases of lessons, and the quality of whole class discussions. Importantly, the IQA was designed to assess student learning opportunities and is consistent with the visions of high-quality instruction that oriented our partner districts' improvement efforts.
- Teachers' and coaches' scores on the Learning Mathematics for Teaching instrument.[28] This paper-and-pencil multiple choice instrument assesses the depth of a specialized form of mathematical knowledge that is specific to teaching, known as mathematical knowledge for teaching (MKT).[29]
- Video-recordings of select pull-out professional development sessions for teachers during the first four years of our partnership work, and video-recordings of pull-out professional development sessions for teachers, coaches, and school leaders during the second four years.

- Audio-recordings of select teacher collaborative meetings during the first four years of our partnership work, and video-recordings of teacher collaborative planning meetings in three or four schools in each of two districts in the second four years.
- On-line assessment of teachers' advice networks completed by all three hundred mathematics teachers in the participating schools, not just the 120 teachers who were full participants. This assessment asked teachers to name the colleagues to whom they turned to for advice about instruction and to indicate how often they sought advice from each.

In addition, the districts provided us with access to mathematics achievement data for students in the participating 120 teachers' classrooms. We achieved almost 100 percent success in collecting all types of data (including the online survey and on-line assessment of teachers' advice networks) in all districts each year.

The retrospective analyses that we conducted of these additional types of data informed the development of a theory of action for instructional improvement at scale and thus the attainment of our research goal. Later in this chapter, we give detailed descriptions of the IQA and MKT measures and also of two additional measures that we created as we used them in many of the retrospective analyses that we conducted.

Formulating recommendations about revising improvement strategies
Before we could make recommendations to the districts about how they might revise their improvement strategies, we had to explain *why* their strategies were being implemented in the ways that we had documented rather than as district leaders had intended.[30] To develop these explanations, we first analyzed the interviews conducted with participants in a particular district to infer how the supports for the learning of members of particular role groups had actually been implemented and to assess whether their practices had changed from the previous year. We then used the interpretive framework described above to explain the changes (or the lack of change) in participants' practices by assessing the learning opportunities provided by the supports as they had actually been implemented. We were then in a position to formulate recommendations to district leaders about how either their improvement strategy or its implementation might be improved. To do so,

we drew on the current iteration of our theory of action for instructional improvement at scale, the final version of which is presented in subsequent chapters of this book.[31]

We shared our findings and recommendations with the leaders of our partner districts each year by writing *District Feedback and Recommendations Reports* of approximately fifteen single-spaced pages. (Sample reports can be downloaded at http://vanderbi.lt/mist.) These reports built directly on the *District Design Documents* by presenting our findings and recommendations for each district's current improvement strategies, thereby focusing directly on the work that district leaders were attempting to accomplish. For each strategy reported in the *District Design Document*, we reiterated the envisioned forms of practice that constituted the goal of the strategy and described the intended supports for the development of the envisioned practices. We then reported our findings about how that strategy was actually being implemented in schools and classrooms, explained why this was the case, and made our recommendations for revising the strategy.

We submitted the reports to district leaders in May, approximately one week before a two-hour meeting scheduled in each district with leaders from departments that were involved in the improvement effort. We requested that these meetings be held in conference rooms and we spoke from notes rather than PowerPoint slides in order to encourage an open discussion of our findings and recommendations. Our role was solely advisory and we viewed our recommendations as proposals that were open to debate. Although our findings were often disappointing for district leaders, they were never defensive and instead engaged in an open dialogue about the current status of the district's improvement efforts and about how those efforts might be improved.

The following October, after district leaders had developed plans for the next school year over the summer, we interviewed them again to document their revised instructional improvement strategies. Our analyses of these interviews indicated that, looking across districts and across years, district leaders acted on an average of 67 percent of our recommendations (see chapter 14). The extent to which district leaders saw value in partnership work and acted on our recommendations is a relatively strong indication that our district partnerships were successful in achieving our pragmatic goal of adding value to the districts' improvement efforts (see the afterword and the appendix).

Research Goal: Developing a Theory of Action for Instructional Improvement at Scale

In addition to supporting our partner districts' improvement efforts by providing feedback and recommendations on the districts' improvement strategies, we addressed our research goal of developing a theory of action for instructional improvement that would be relevant to other districts aiming at rigorous goals for all students' learning. The question we asked ourselves was: What empirically grounded recommendations would we make to other districts based on what we have learned while collaborating with our four partner districts?

Initial conjectures about productive improvement strategies

In preparing to work with our district partners, we developed a set of initial conjectures about productive improvement strategies by drawing on then current literature in mathematics education, teacher education, educational policy, and leadership.[32] We viewed these initial conjectures as provisional and quite possibly wrong, and tested and revised them throughout our collaborations with the four districts. With hindsight, it is clear that our initial conjectures were far too global and under-specified. As a consequence, we had to claw our way up to what we came to call the level of concrete practice. The types of additional questions that we needed to answer included: In what types of activities should coaches engage teachers when they facilitate teacher collaborative time and when they work with them one-on-one in their classrooms?, What do coaches need to know and be able to do to implement these activities productively?, and What specific leadership practices do principals need to develop in order to create conditions in their schools that support teachers' development of ambitious and equitable instructional practices? In addition, it soon became apparent that we had initially failed to consider several essential aspects of a coherent instructional system, the most glaring of which was the provision of supplemental supports for currently struggling students.

The process of testing and revising our conjectures was informed by new studies reported in the research literature, the annual data collection, analysis, and feedback cycles, and retrospective analyses that looked across years and that drew on the multiple types of data that we collected each year.

Research literature

We have noted that current relevant research that can inform the design of instructional improvement strategies becomes increasingly thin the further one moves away from the classroom. Nonetheless, findings reported in the literature have, on occasion, provided evidence for the revision of our conjectures. For example, a study of instructional coaching reported by Cynthia Coburn and Jennifer Lin Russell found that the issues on which coaches press teachers can influence the questions that teachers ask each other when the coach is not present.[33] This finding indicates that in addition to supporting teachers' learning directly, effective coaching can also support more productive teacher advice networks. We revised the theory of action in light of this finding and conducted analyses which showed both that teacher advice networks can be an important support for instructional improvement and that effective teacher collaborative meetings can support more productive teacher networks (see chapter 8).

Annual data collection, analysis, and feedback cycles

The data collection, analysis, and feedback cycles that we conducted in partnership with district leaders each year gave us two types of opportunities to test and revise our conjectures about productive improvement strategies. First, the leaders of the four districts attempted to implement a relatively high proportion of our recommendations for revising their improvement strategies. We therefore had the opportunity to investigate the consequences of these recommendations in subsequent years, and thus the conjectures on which they were based.

Second, in formulating recommendations to our partner districts, we had to propose actionable ways of addressing the problems that each district was encountering while taking account of the districts' current capacity for instructional improvement. To be clear, our focus when formulating recommendations to districts was pragmatic given that the recommendations could be consequential for the mathematics learning of a large number of students. However, once we completed each annual data collection, analysis, and feedback cycle, we stepped back and framed each district as a case of supporting teachers' development of ambitious and equitable instructional practices at scale. In doing so, we considered whether any of our recommendations to a particular district had implications for revising or elaborating our current conjectures, and if they did whether they might have more general implications. Working with four districts was helpful as we could first determine whether a recommendation made to one district might be relevant for one or more of the other districts, and could clarify conditions under which the recommendation would be relevant more generally.

This process of revising conjectures is represented in figure 2.1. As the figure shows, the development of actionable recommendation to the districts

FIGURE 2.1 **Relationship between the pragmatic and research-oriented levels of analysis**

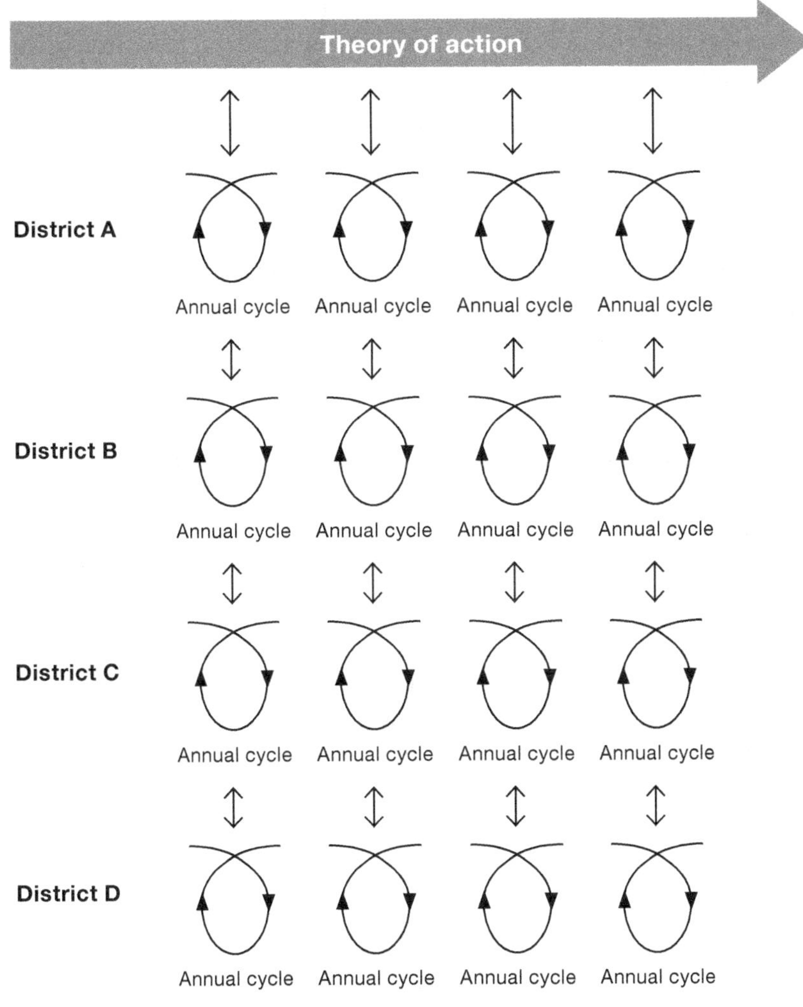

(pragmatic goal) and the ongoing revision of the theory of action were interdependent. On the one hand, the insights that we developed while formulating recommendations to the districts informed the revision of our theory of action. On the other hand, the current iteration of the theory of action informed our formulation of recommendations to the districts about how they might revise their improvement strategies.

If our goals had been purely pragmatic and we had conducted annual data collection, analysis, and feedback cycles in one or more districts to support their instructional improvement efforts, we would have engaged in action research, though at the district rather than the classroom level. However, as the figure makes clear, we also aimed to contribute to the field's understanding of what it takes to support instructional improvement at scale. This concern for both pragmatic relevance and generalizability is a key characteristic of design research studies.[34] The design research methodology is appropriate when there are significant gaps in the relevant research and when it is difficult, if not impossible, to identify cases where the forms of learning and improvement of interest are occurring in situ and can be investigated by conducting observational studies.[35] As we have noted, the research base on which we could draw was limited and it would have been difficult, if not impossible, to identify in advance large urban districts that were likely to make significant progress in supporting teachers' development of ambitious and equitable instructional practices. We therefore required an interventionist methodology in which we partnered with districts to bring about the phenomenon we wanted to study, the implementation of improvement strategies that we conjectured would be productive. In the process, we became co-designers of improvement strategies with district leaders and conducted four parallel design studies at the level of a large district.[36] Partnerships of this type between researchers and practitioners that employ design research methods to investigate and support the development of specific instructional practices in particular content areas are typically called Design Research Partnerships.[37]

Retrospective analyses

Earlier in this chapter, we described the different types of data that we collected in each district each year, and clarified that the districts provided us with access to student achievement data from state assessments each year. We did not begin conducting retrospective analyses of these multiple types of data until the third year of our partnerships with the districts. During the first two years, we developed the method we have described for analyzing the two hundred participant interviews in order to provide the districts with timely feedback in their improvement strategies. During this time, we also developed the interpretive framework that we used when formulating recommendations to the districts on how they might revise their improvement strategies to make them more effective.

In addition, as part of our preparation for working with the districts, we looked for appropriate measures for the aspects of teachers' knowledge, perspectives, and practice that we wanted to assess. We identified two existing measures that were appropriate for our needs, measures of the quality of teachers' instructional practices and of their mathematical knowledge for teaching. However, there were no measures of teachers' visions of high-quality mathematics instruction or of their views of their students' current mathematical capabilities. We therefore constructed these two measures during the first two years of partnership work and subsequently used them and the two existing measures in many of the retrospective analyses that we report in this book. We therefore first describe the two pre-existing measures, the Instructional Quality Assessment Mathematics Toolkit and Mathematical Knowledge for Teaching, followed by descriptions of the two measures that we created, Vision of High-Quality Mathematics Instruction and Views of Students' Mathematical Capabilities.[38]

Instructional Quality Assessment (IQA). Given that we wanted to assess the quality of middle-grades mathematics teachers' instructional practices across multiple years, we chose a measure that was aligned with the vision of ambitious and equitable instruction that oriented our partner districts' improvement efforts. Further, because three of the four of the districts had adopted an inquiry-oriented textbook series, CMP2, and the fourth had supplied CMP2 to schools as a supplemental resource, we selected a tool that was aligned with the Launch-Explore-Summarize format of CMP2 lessons.[39] The Launch phase of a lesson includes the introduction of a cognitively demanding task. This is followed by the Explore phase which is when students work (often in groups) to solve the task. Finally, the Summarize phase includes the whole-class discussion in which students share and compare their strategies to further develop their mathematical understanding and reasoning.

We found that the IQA was a good fit for assessing these aspects of instructional quality. One set of rubrics assesses the cognitive demand of classroom activity over the course of the lesson, including a measure of the rigor of the tasks used in the lesson (Task Potential), a measure of the rigor of the tasks as they are actually implemented in a lesson (Task Implementation), and a measure of the quality of the whole class discussion that the teacher leads after students have had a chance to work on tasks (Discussion). Additional rubrics focus on specific aspects of discourse during the whole-class discussion, including the percentage of students participating in the discussion (Participation), the extent to which the teacher and students link, or connect, students' contributions to one another (Teacher Linking and Student Linking, respectively), the extent to which the teacher asks students to provide evidence for their claims (Teacher Press), and the extent to which the students provide evidence for their claims (Student Providing). Scores on each of these rubrics range from zero to four with scores at the low end representing a lack of (mathematical) activity, scores in the middle representing procedural activity, and scores at the high end representing conceptually rich classroom activity.[40] Given the districts' focus on instruction that was equitable as well as ambitious, we developed several additional rubrics to assess the introduction or launch of tasks, described in chapter 3, to complement the existing IQA rubrics.[41]

Mathematical Knowledge for Teaching (MKT). The findings of a number of studies indicate that it is essential to distinguish between general mathematical content knowledge assessed, for example, by the SAT, and content knowledge that is specific to mathematics teaching.[42] For example, mathematics teachers need not to only understand the meaning of fractions, they also need to understand how students tend to make sense of fractions and their common misunderstandings, and know to represent key ideas of fractions to students. Heather Hill, Deborah Ball, Hyman Bass, and their colleagues call this kind of knowledge *mathematical knowledge for teaching* (MKT) to distinguish it from the mathematical knowledge needed in other professions (e.g., nursing, engineering).[43] We used a multiple choice, pencil-and-paper instrument created by Hill, Ball, Bass, and colleagues to assess teachers' and coaches' MKT on two dimensions specific to the middle grades: number concepts and operations, and patterns, functions, and algebra.[44]

Vision of High-Quality Mathematics Instruction (VHQMI). From the outset, one of MIST's guiding conjectures was that improvement in teachers' instructional practice would be more likely in schools and districts in which teachers and instructional leaders shared a vision for how best to equitably support students' mathematics learning.[45] Importantly, an instructional vision is aspirational and represents an improvement goal that teachers and instructional leaders are working toward, thus giving a sense of direction to collective efforts to improve the quality of instruction. We also conjectured that shared visions might be negotiated over time and that in the process teachers, coaches, school leaders, and district leaders would develop a common language for describing high-quality mathematics teaching and learning.[46]

In order to assess the extent to which individuals' instructional visions were compatible with each other and with the vision promoted by district leaders (as well as whether they improved over time), we developed an instrument for assessing participants' *visions of high-quality mathematics instruction* (VHQMI).[47] The interviews conducted each year with teachers, coaches, school leaders, and district leaders included questions in which we asked interviewees to imagine that they were observing a teacher's mathematics instruction for one or more lessons, and to indicate how they would determine whether the instruction was of high quality. Follow-up probes then pressed interviewees on what the nature of the tasks, classroom discourse, and the teacher's role would look like if the lesson was of high quality.

The VHQMI instrument that we created to code interviewees' response to this series of questions consists of separate rubrics that focus on mathematical tasks, classroom discourse, the role of the teacher, and student engagement in classroom activity. The rubrics have four or five levels and map roughly onto the IQA rubrics described above. In general, the highest levels align with recent research on mathematics classrooms and student learning, emphasizing teachers supporting students as they attempt to complete challenging mathematical tasks and make substantial contributions to classroom mathematics discussions. The lowest levels of the rubrics align with conventional instruction, emphasizing clarity in the teacher's presentation of content, students being on-task, and mastering the basics before moving to problem solving. The intervening levels represent a transition in participants' characterizations of high-quality instruction, initially taking up only new *forms* for describing classroom practice (e.g., "group work," "discovery," "hands-on" tasks) and eventually also articulating the intended *functions* of those new forms in supporting students' learning. This distinction between the form and the function of particular aspects of instruction has been used in education research to describe the extent to which practitioners take up reform-oriented concepts and practices.[48] Research suggests that when practitioners initially attempt to improve instruction, they may implement the intended forms of instruction without necessarily adopting the intended function.[49] For example, participants might initially stress the importance of eliciting student thinking, stating that it is important to get all students engaged, rather than the intended function—that student thinking should be used to drive instruction. In the case of using the VHQMI instrument, interviewees were asked to explain *why* the aspects of instruction they identified were indicative of high-quality mathematics instruction in order to assess the function that they expected a certain form to serve.

Views of Students' Mathematical Capabilities (VSMC). Based on prior research, we anticipated that the extent to which teachers in our partner districts developed ambitious and equitable teaching practices would be influenced by their views of their students' current mathematical capabilities.[50] Furthermore, we anticipated that this might be especially the case when teaching students from historically marginalized groups; existing research shows that mathematics teachers often justify engaging students from non-dominant groups solely in procedurally oriented activities in terms of their perceptions of these students' capabilities.[51] We searched for existing tools for assessing teachers' views of their students' current capabilities that could be used on a large scale, but were unable to locate any that took into account a specific vision of mathematics instruction. Thus, similar to the assessment of instructional vision described above, we developed an interview-based assessment specific to eliciting teachers' views of their students' current mathematical capabilities (VSMC).

Based on a review of the literature and our initial interviews with teachers in our partner districts, we conceptualized VSMC along two dimensions: a) teachers' *explanations* regarding why their students face difficulty in mathematics; and b) teachers' descriptions of the *adjustments* they make to instruction for students they perceive as facing difficulty.[52] Together, teachers' explanations of students' difficulty and their description of adjustments provide insight into teachers' views of their current students' mathematical capabilities with respect to an ambitious and equitable vision of instruction.

To assess teachers' explanations regarding why their students faced difficulty, we asked questions like, "When your students don't learn as expected, what do you find are typically the reasons?" To assess teachers' descriptions of the adjustments they made to instruction for students they perceived as facing difficulty, we asked teachers about what they did to support those students, and about their rationales for making particular adjustments. In addition, we asked a related set of questions to other role groups (i.e., coaches, school leaders, district leaders). For example, to elicit school leaders' and coaches' explanations, we asked, "When students in your school don't learn as expected, what do you find are typically the reasons?" And, we asked about coaches' and school and district leaders' *expectations* for how teachers should adjust their instruction to support different groups of students facing difficulty in mathematics.

We then designed a set of rubrics to code interviewees' responses specific to the two dimensions of VSMC. Across both dimensions, we use the language of *productive* and *unproductive* to signal perspectives that influence whether teachers take action to engage all students in rigorous activity.[53] One rubric focuses on the nature of explanations for students' difficulty. Productive explanations are ones that attribute student difficulty to instructional and/ or schooling opportunities, whereas unproductive explanations are ones that attribute student difficulty to inherent traits of the student, or their family or community. A second rubric focuses on the adjustments teachers made (or are expected to make) to support students perceived as facing difficulty. Productive adjustments aim to enable students who face difficulty to participate in rigorous mathematical activity. For example, teachers described carefully planning for the introduction of a cognitively demanding task, with the goal of ensuring all students could begin to work the task.[54] Another example of a productive adjustment entails engaging students in explicit conversation regarding how to construct mathematical explanations and justifications.[55] On the other hand, unproductive adjustments aim to reduce the rigor of the learning goals for students currently encountering difficulty. For example, teachers often described "spoon-feeding" students the discrete steps to solve a cognitively demanding task, or deciding to only focus on the mastery of basic skills with particular students.

Summary. It was with these measures in hand that we began conducting retrospective analyses in the third year of partnership work. We organized the project team into a number of analysis teams with overlapping membership that focused on conjectures relating to the various elements of the coherent instructional system and to school instructional leadership and district instructional leadership. The findings of their analyses informed the ongoing revision of the conjectures about productive improvement strategies and thus our Theory of Action. This was a genuine mixed methods investigation as the analysis teams used a range of qualitative and quantitative methods depending on the questions they were addressing. Members of the analysis teams contributed to the chapters that follow, in which we discuss the findings of our own and others' work, describe takeaways for school and district instructional leaders, and suggest next steps for research that aims to inform school and district instructional improvement efforts.

THREE

Specifying Goals for Students' Mathematics Learning and the Development of Teachers' Knowledge, Perspectives, and Practice

KARA JACKSON, ANNE GARRISON WILHELM, AND CHARLES MUNTER

AS DESCRIBED IN CHAPTER 1, a coherent instructional system must be oriented by a set of learning goals for students, and a corresponding vision of what counts as high-quality mathematics instruction. In this chapter, we first elaborate on the goals for students' mathematical learning and the vision of high-quality mathematics instruction that grounded and oriented the joint work between the partner districts and the research team.[1] We then describe a vision of *ambitious* and *equitable* teaching that enables all students, including historically disadvantaged groups of students, to achieve such goals. Next, we share findings from the MIST study that suggest that teachers need more than deep content knowledge if they are to develop ambitious and equitable teaching practices; they also need to develop sophisticated instructional visions and to view their students as capable of engaging in rigorous mathematical activity. As we discuss below, these findings have considerable implications for instructional leaders charged with designing and enacting professional learning supports in districts pursuing reform at scale.

RIGOROUS GOALS FOR STUDENTS' MATHEMATICAL LEARNING

Mathematical learning goals for students reflect a set of values regarding what is worth knowing and doing mathematically. In the case of our partner districts, they adopted a set of goals for students' learning that were broadly compatible with those represented in the *Common Core State Standards—Mathematics* and the National Council of Teachers of Mathematics' (NCTM) *Standards*.[2] For example, goals included developing procedural fluency and conceptual understanding of key mathematical ideas in a range of domains. These goals indicated that district leaders privileged students making sense of problem situations, and knowing *why* it makes sense to use particular procedures in certain situations. In addition, district leaders argued it was important that students develop the ability to communicate effectively about mathematics. Our district partners described how these kinds of goals, as opposed to goals that focused solely on developing skills to solve routine sets of problems, would prepare students for life beyond secondary schooling, including postsecondary education and careers.

Moreover, all of our partner districts articulated a clear commitment to ensuring that students from historically disadvantaged populations would have access to high-quality instruction and would develop rigorous understandings of mathematics and productive mathematical dispositions. In our partner districts, such groups predominantly included African American and Latinx students, students identified as English learners, and students identified as receiving special education services. Central office leaders in each of the districts recognized the historical inequities in opportunities that had been provided to these populations of students and knew they had to do better.

As described in chapter 2, the fact that the districts adopted these goals at the start of the study is noteworthy. It was 2007, with high-stakes accountability pressures in full force in the United States. Our partner districts were under immense pressure to show improvement in students' performance—especially for the student populations mentioned above. Moreover, the assessments were predominantly focused on students showing proficiency with using skills to solve routine sets of problems. As such, the districts' goals for students' learning were above and beyond the goals mandated and assessed by the states in which they were located. However, over the course of our partnerships, the *Common Core State Standards* were developed and adopted in two of the three states in which our partner districts were located, and although the third state did not adopt the *Common Core State Standards*, it implemented a more rigorous set of state standards for mathematics and an associated assessment. In this way, our partner districts' articulation of learning goals for students (and vision of instruction, as described below) anticipated the increase in rigor in standards for mathematics teaching and learning marked by the transition to Common Core.

A VISION OF AMBITIOUS AND EQUITABLE MATHEMATICS INSTRUCTION

Against these goals for students' learning, our partner districts promoted a vision of what we have come to refer to as ambitious and equitable mathematics instruction. Visions of instruction specify what should happen between teachers and students in order to achieve the intended learning goals for students—which, in the case of our partner districts, concerned the rigorous learning goals described above. An ambitious and equitable vision, which we will elaborate below, is compatible with the vision promoted by the NCTM, from its original *Standards* to the more recent *Principles to Actions*.[3] The vision is grounded in a robust research base that has solidified over the past twenty-five years or so in the mathematics education and learning sciences research communities regarding concrete forms of instructional practice necessary to support students to achieve rigorous learning goals.

That said, whereas there has been increasing focus given to equity concerns in mathematics education research, it became clear in our initial work with our partner districts that the research base regarding how to concretely support historically marginalized groups of students to participate substantially in rigorous mathematical activity, especially in the middle-grades, was thin.[4] As such, early in the MIST project, we initiated a line of research aimed explicitly at further specifying forms of teaching practice that appear to support historically marginalized groups of students.[5] As this research evolved, we were in a position to support our partner districts to incorporate equity-specific forms of practice into their vision, and thus their professional learning strategies. We describe this elaborated vision as ambitious and equitable to signal that this vision poses significant learning demands for most teachers and that it is centrally concerned with supporting historically marginalized students to develop robust understandings of mathematics and productive mathematical dispositions.[6] Here, we briefly describe this vision.

A central aspect of an ambitious and equitable vision concerns the *nature of the tasks*, or problems, students are expected to solve. A key distinction concerns whether the task is of low- or of high-cognitive demand, or challenge.[7] Low-cognitive demand tasks ask students to apply known procedures to predictable sets of problems. For example, the task in figure 3.1a tells students to graph a pair of equations and to identify where the lines intersect. There is little ambiguity in how to solve such tasks; the teacher or textbook is positioned as having mathematical authority, and the students' role is to mimic what has been shown to them. On the other hand, high-cognitive demand tasks are often open-ended, in that they can be solved in

FIGURE 3.1 Examples of tasks with different levels of cognitive challenge*

For exercises 1-2: a. Graph each pair of equations on the same coordinate grid. b. Identify the point of intersection. 1. y = 12x + 60 y = 15x 2. x + y = 10 y = 2x + 4	Two parking garages close to where Michael works charge the following rates for each month of parking: *Park-in-lot:* • Maintenance fee of $60. • $12 per day. *City Parking:* • No maintenance fee. • $15 per day. Which parking garage should Michael choose? a. Show all of your mathematical work. b. Based on your work, make a recommendation to Michael. c. What information did you consider in making your decision? What additional information did you want (if any)?
A	B

*This is a modified version of Figure 1, p. 6 published in Melissa Boston and Anne Garrison Wilhelm, "Middle School Mathematics Instruction in Instructionally Focused Urban Districts," *Urban Education* 52, no. 7 (2017): 829–861. doi:10.1177/0042085915574528.

multiple ways, and expect students to engage in mathematical reasoning. For example, the task in figure 3.1b expects students to model the costs of the parking garages, and to do so, students need to assess the situation and make decisions regarding what information is relevant, and how to represent it. As illustrated in the sample task, tasks of high-cognitive demand often require that students explain or justify their solution strategies.

Research has documented that when students are primarily asked to solve tasks of low-cognitive demand, they have few opportunities to develop an understanding of *why* particular procedures are appropriate in certain problem situations or why they work. In addition, they have few opportunities to develop disciplinary practices, like problem-solving skills or the ability to explain their mathematical thinking.[8] On the other hand, when students are provided access to instruction that regularly features tasks of high-cognitive demand, students tend to develop conceptual understandings of mathematical ideas, are more flexible problem-solvers, and view mathematics as a sense-making activity.[9]

That said, research also suggests teachers find it challenging to *maintain the cognitive demand of a task*.[10] In fact, over the first four years of the MIST project, although most teachers began lessons with cognitively demanding tasks, they often lowered the challenge of the task over the course of the lesson by suggesting a method of solving the task.[11] Doing so reduces the opportunities that students have to develop robust understanding of the focal mathematical ideas. We also observed that teachers often appeared to do this in response to students' difficulties in getting started on solving tasks—and that often the students who had difficulty getting started were students from historically marginalized groups.

In response, we elaborated a key practice of ambitious and equitable instruction—*introducing cognitively demanding tasks* so that all students are able to get started on solving these kinds of tasks in productive ways.[12] Our analysis suggested that in the introductory phase of a lesson (often referred to as the launch), it was important that teachers supported students to develop common language specific to key contextual features of a problem-solving scenario *and* to key mathematical ideas that were to be explored in the task.[13] Key contextual features refer to those aspects of the scenario that are likely to be unfamiliar to some students, yet really matter in deciding what steps to take in solving the problem. For example, in figure 3.1b, in order to mathematically model the Park-in-lot scenario, students need to understand that parking garages offer monthly parking, and what a "maintenance fee" is, especially that it is a fee you pay once, not every day. Key mathematical ideas refer to the ideas that students are asked to represent, or make sense of, in the task at hand. In the focal example, students might use tables, graphs, and/or equations to represent the accumulation of money over time in the two different garages. In order to do so, it is essential that students understand that money accumulates, on a daily basis, in light of how many days a person parks their car. Moreover, based on our analysis, it was not enough for the teacher to clarify key contextual features and mathematical ideas associated with the task at hand. Instead, it proved critical that students had an opportunity to develop *common language* regarding the key contextual features and mathematical ideas; common language supported students to make sense of solution strategies together in small groups and in whole-class discussions.

It is worth noting that teachers sometimes might choose to alter the scenarios suggested in curriculum materials, like the parking garage scenario in figure 3.1b, so that the task is more relevant to their students' lives. However, we found in our partner districts that even in cases in which teachers purposefully altered the scenario, it was still important that teachers provided students opportunities to develop common language in the launch to describe the contextual features and the key mathematical ideas. Without deliberate attention to developing common language, even when scenarios were assumed to be familiar, some students still struggled to make sense of the problem situation and the mathematical ideas they were expected to represent.

Another central aspect of a vision of ambitious and equitable instruction concerns the role of *discourse*, or mathematical conversations, students are expected and supported to have. Student thinking is at the heart of ambitious and equitable teaching; an assumption of such a vision is that all students have worthwhile ideas on which to build, and that a primary goal of teaching is to find out what students currently know, and build instruction based on this knowledge in order to advance their mathematical understandings.[14]

Research has identified the critical importance of both *small-group and whole-class discussions* in advancing students' conceptual understanding and development of mathematical dispositions. In small groups, it is important that students share and refine their thinking and that teachers monitor students' thinking in order to plan for productive whole-class discussions. In whole-class discussions, it is important that students reveal and compare their strategies in order to make connections between them and to build shared understandings of key mathematical ideas.[15] However, the extent to which small-group and whole-class discussions are beneficial depends crucially on the nature of the task that students are discussing, and the role of the teacher. As described by Mary Kay Stein and Margaret Smith, "[P]roductive discussions that highlight key mathematical ideas are unlikely to occur if the task on which students are working requires limited thinking and reasoning."[16]

Assuming students are working on solving and discussing cognitively demanding tasks, the teacher plays a crucial role in supporting all students to participate substantially and learn through discussion in the classroom. One aspect of the teacher's role concerns *establishing norms* that guide interactions in the small-group and whole-class setting.[17] Important norms to establish include the importance of making thinking public, including the value in learning from "mistakes," and of monitoring that no one student dominates discussion at the expense of others.[18] In addition, it is important to establish norms regarding what counts as an acceptable mathematical explanation.[19] In particular, explanations that include descriptions of both the steps taken to solve a problem *and* the rationale for taking those steps are much more likely to support students' learning than explanations that focus on steps alone.

Another critical aspect of the role of the teacher concerns *pressing students to elaborate their reasoning and to make connections between their peers' solutions and key mathematical ideas.*[20] This is especially crucial in whole-class discussions. It was quite common in our partner districts for whole-class discussions to take the form of "show-and-tell," in which a student or students shared their solution strategies to a challenging task, while other students listened politely.[21] Although show-and-tell forms of discussions can be an important step in teachers' developing practice, on their own, they are unlikely to advance students' learning. Instead, it is essential that teachers press students to make sense of others' solutions in relation to key mathematical ideas. This entails pressing the presenting students to provide reasoning, or the rationale implicit in their solution strategy, and pressing the listening students to make sense of that reasoning. In addition, students' learning is advanced by pressing all students to make connections between solutions that highlight underlying mathematical ideas.

In addition to pressing students to provide reasoning, or to make connections, it is also crucial that a teacher *supports students to learn how to engage in such forms of discourse.*[22] MIST researchers conducted an analysis of classrooms in which African American students performed better than would be predicted by their previous two years of assessment scores, and in which there was evidence that teachers were posing cognitively demanding tasks and orchestrating whole-class discussions in which students were pressed to explain their reasoning.[23] The analysis revealed seven forms of practice that characterized classroom instruction. One form of practice concerned *coaching students*, in which teachers explicitly intervened to support a student or group of students to meet their expectations. Coaching occurred during both small-group and whole-class discussions. For example, in both settings, the teacher would remind students of how to ask questions of one another that focused on sense-making, or on making connections between solutions. Another form of practice that often occurred during small-group discussions entailed *attributing mathematical authority to students.* Within these classrooms, teachers often responded to student requests for assistance that got at the core of the mathematical task by encouraging them to either talk to their peers or to think further about the particular question. This served to both position the students as competent and to maintain the cognitive demand of the task.[24]

ASSESSING TEACHER KNOWLEDGE, PERSPECTIVES, AND PRACTICE

Above, we described the vision of ambitious and equitable instruction that our partner districts adopted, and that research suggests is necessary to ensure that students are afforded high-quality, empowering learning opportunities in the mathematics classroom. Such a vision of high-quality mathematics instruction represents a set of *learning goals for teachers.* That is, teachers need to learn to enact the forms of practice specified above. For most teachers, doing so represents a significant change from their current practice. In most US mathematics classrooms, students are taught procedures for solving tasks of low-cognitive demand, and discussions focus on sharing correct answers, with little room for discussion of reasoning.[25] This presents a substantial challenge to districts attempting to achieve a vision of ambitious and equitable instruction.

A key aspect of figuring out how to support significant teacher learning entails knowing what achieving such a vision requires on the part of teachers' knowledge and perspectives. This surfaced as an important consideration in our collaboration with our partner districts, and as such, we carried out a line of inquiry that focused on answering questions that stem from it. For example, what is required for teachers to orchestrate productive interactions among students, or to foster the development of students' positive, productive relationship to and understanding of mathematics? What must teachers' own relationships with and knowledge of mathematics look like if they are to be effective in supporting students to develop rigorous understandings of mathematics? How do teachers' views of their students' capabilities impact the forms of teaching they implement?

Answers to questions like these have substantial implications for those tasked in districts and schools with supporting teacher learning; they need to determine starting points and associated learning demands with respect to these and other facets of teachers' practice. In our work as researchers, answering such questions required having various forms of assessments. Specifically, we wanted to be able to assess and document changes in a) what classroom teaching looked like, b) teachers' mathematical knowledge, c) teachers' instructional visions, including how they talked about and reflected on their practice, and d) teachers' views of their students. There were existing measures of the quality of teachers' instructional practices and mathematical knowledge for teaching. However, there were no measures of teachers' instructional vision or of their views of their students' mathematical capabilities. In chapter 2, we describe the two pre-existing measures that we decided to use, the Instructional Quality Assessment Mathematics Toolkit and Mathematical Knowledge for Teaching, followed by descriptions of the measures that we created ourselves, Vision of High-Quality Mathematics Instruction and Views of Students' Mathematical Capabilities.[26] In the next section, we briefly describe the MIST sample of teachers with respect to each of the forms of teachers' practice, knowledge, and perspectives; this sets the stage for a subsequent section that elaborates on the *relations* between teachers' knowledge, perspectives, and practice. Note that our use of the term "perspectives" refers to teachers' instructional visions and their views of their students' mathematical capabilities.

MIST Sample of Teachers' Classroom Practice

In the first year of the study, across the four districts, we found that a high percentage of teachers' lessons featured cognitively demanding tasks. However, the cognitive demand of the tasks declined across the course of the lesson.[27] Teachers often suggested procedures to solve the tasks, thereby taking away opportunities for students to reason themselves about important mathematical ideas. Further, the majority of discussions consisted of students demonstrating procedures in a show-and-tell format or providing brief responses to teachers' questions. However, in one district, almost half of lessons involved teachers and students engaged in conceptually-focused whole-class discussions.

In investigating how teachers introduced tasks in years three and four of the MIST study, we found that lessons with higher quality launches (i.e., launches in which students developed common language to describe key contextual features and mathematical relationships, and in which the cognitive demand of the task was maintained) tended to have higher quality discussions.[28] However, launches in which the teacher and students developed common language to describe the key contextual features and mathematical relationships and maintained the cognitive demand of the task were quite rare (6.7 percent of our sample).[29]

Comparisons with other large studies suggest that the MIST sample exceeded the instructional quality of nationally representative samples and was on par with other instructionally focused urban middle schools.[30] Over the course of the study, on average, instructional quality remained fairly stable, with minimal changes.[31] Importantly, there were significant differences between districts, and in the following chapters we describe how different features of the instructional system may have contributed to these differences.

MIST Sample of Teachers' Mathematical Knowledge for Teaching

Mathematics education researchers have specified the importance of distinguishing between common content knowledge, or understandings of mathematical ideas, and content knowledge that is specific to mathematics teaching.[32] For example, mathematics teachers need to not only understand the meaning of fractions, they also need to understand how students tend to make sense of fractions and their common misunderstandings, as well as how to represent key ideas of fractions to students. This kind of knowledge – what Heather Hill, Deborah Ball, and colleagues refer to as *mathematical knowledge for teaching* (MKT) – is distinct from the mathematical knowledge needed in other professions (e.g., nursing, engineering).[33]

Across our sample, we found that the average teacher had developed MKT on par with that of the average middle school mathematics teacher in the United States.[34] However, there was significant variation by district.[35] Therefore, in some of the partner districts, district leaders needed to do significant work to support teachers' understanding of the content they were teaching, how to represent key mathematical ideas to students, and how to interpret students' understandings of such ideas.

MIST Sample of Teachers' Vision of High-Quality Mathematics Instruction

From the outset, one of MIST's guiding conjectures was that improvement in teachers' instructional practice would be more likely in settings in which teachers and instructional leaders shared a vision for how best to equitably support students' mathematics learning.[36] Importantly, a vision represents something that teachers are working toward–a conception of the instructional practice that they *aspire* to enact. Our assumption was that such shared visions might emerge over time as teachers and leaders developed a common language for describing high-quality classroom mathematics teaching and learning, and that developing such a shared vision would guide collective efforts to improve teaching.

We found that over the first three years of the MIST project, instructional visions consistently improved over time, but patterns and rates of change differed by role group (e.g., teachers, principals, district leaders).[37] The most sophisticated visions were typically articulated by district mathematics specialists and by math coaches, particularly early in the improvement efforts. This pattern aligned with the professional development models that most districts had adopted, in which leaders and coaches participated first and then led sessions for teachers and principals. Instructional vision trends also differed across districts. For example, teachers' visions of what counted as high-quality mathematical tasks were consistently more sophisticated in the district with the longest history of adopting and using reform-oriented mathematics textbooks; this district had been using the first edition of the

Connected Mathematics Project materials for approximately ten years prior to the adoption of CMP2 in the second year of the MIST study.[38]

Given that changes over time in instructional vision were based, in part, on changes in personnel, we also examined a subset of participants for whom we had scores in each of the first four years and found that their instructional visions improved over time.[39] For example, over the first four years of the study, on average, teachers improved in their vision of the role of discussion in supporting students' learning. Whereas early on, on average, teachers suggested that in high-quality instruction, discussion should be limited to small group settings, by the end of four years, on average, teachers suggested that there should also be whole-class discussion following mathematical work in which students have just engaged.

MIST Sample of Teachers' Views of Students' Mathematical Capabilities

In light of existing research, we anticipated that the extent to which teachers in our partner districts developed ambitious and equitable teaching practices would be shaped, at least in part, by their views of their students' mathematical capabilities.[40] Furthermore, we anticipated that this might be especially the case with respect to the teaching of historically marginalized students; existing research shows that mathematics teachers often justify engaging students from non-dominant groups solely in procedurally oriented activities in terms of their perceptions of students' capabilities.[41]

In an analysis of year five data, we found that less than one-fifth of teachers attributed student difficulty to instructional and/or schooling opportunities, and nearly 30 percent of the teachers in our sample attributed students' difficulty *solely* to inherent traits of the students, and/or deficits in their families and communities, thereby locating the responsibility for those students' learning as outside of their purview.[42] Similarly, less than one-fifth of teachers described making productive adjustments to their instruction to support students they perceived as struggling.[43] The majority of teachers described lowering the cognitive demand of activities (e.g., showing students how to solve the problem) if they perceived students were facing difficulty.[44] Moreover, even when teachers viewed students' difficulty as related to instructional opportunities, and thus within their purview, they did not necessarily describe responding to those difficulties in ways that would enable students to participate substantially in rigorous mathematical activity.[45]

Looking across all eight years of the study, while there were some fluctuations from one year to the next for a given district, on average, teachers' views of mathematical capabilities were similar to those described in year five. However, just as with our findings for teachers' instructional vision, in the district that had a longer history of using a reform-oriented curriculum, teachers tended to articulate significantly more productive views of their students' mathematical capabilities. This particular district had invested in providing teachers with professional development specific to developing productive views of their students and knowing how to enact supports that would maintain the cognitive demand of challenging tasks prior to the start of the MIST study.

These findings suggest that most of our partner districts faced a significant challenge in terms of what teachers viewed their students as capable of mathematically, and in terms of the extent to which teachers were prepared to support students perceived as struggling to engage in activity aimed at developing deep understandings of mathematics and productive mathematical dispositions.

RELATIONS BETWEEN ASPECTS OF TEACHERS' KNOWLEDGE, PERSPECTIVES, AND PRACTICE

Above, we reported on the research team's assessments of MIST sample teachers' knowledge, perspectives, and practice. Whereas each of those assessments provides useful information, we were particularly interested in the *relations between* aspects of teachers' knowledge and perspectives, and the development of ambitious and equitable instructional practices. Figuring out how teachers' MKT, instructional visions, and views of their students' capabilities matter for the forms of teaching they develop supports district and school instructional leaders to be much more intentional in designing professional learning supports to advance the quality of classroom instruction.

Mathematical Knowledge for Teaching

A substantial body of mathematics education research reveals how MKT influences teaching. For example, there is considerable evidence that MKT plays a role in teachers' task selection and in the extent to which they maintain the cognitive challenge of tasks throughout a lesson.[46] This was also the case in MIST analyses.[47]

However, while MKT clearly matters, teachers' perspectives play a role in *how* that knowledge influences their instructional practice.[48] For example, we compared the teaching practices of teachers with the least developed MKT with the practices of teachers with more sophisticated MKT, and found that MKT alone was not responsible for more sophisticated teaching practices; instructional vision also mattered.[49] For example, teachers with more sophisticated MKT were more likely to select cognitively demanding tasks if they also articulated a sophisticated vision of teaching. This makes sense as sophisticated visions indicate that teachers value cognitively demanding tasks. Similarly, other research has found that teachers' perspectives about mathematics and teaching and learning, and their goals for student learning, influence the relation between MKT and instructional practice.[50]

Findings such as these suggest that whereas teachers' MKT is clearly important, improving MKT alone will not result in instructional improvement. Within the MIST project, it became increasingly clear that teachers' instructional visions and their views of their students' mathematical capabilities shaped their practice in significant ways. In what follows, we elaborate on how teachers' perspectives related to teachers' instructional practice within our sample. To understand the contribution of teachers' instructional visions and views of their students over and above that of MKT, in each of the analyses reported below, we controlled for teachers' MKT. In other words, the relations we report factor out MKT by holding it constant.

Instructional Vision

Within the MIST project, we found that teachers' instructional visions were significantly related to their task selection and to the maintenance of the cognitive demand.[51] That is, teachers who provided more sophisticated descriptions of high-quality mathematics teaching—and descriptions of the role of the teacher in particular—were more likely to use mathematically rich tasks with students, and in ways that maintained the challenge of the tasks throughout the lesson. In contrast, we found that teachers' instructional visions were not significantly related to the quality of students' participation in whole-class discussion.[52] In other words, for teachers in our sample, articulating a sophisticated instructional vision was not associated with greater learning opportunities for students within the whole-class discussion, over and above the influence of the cognitive demand of the task. Together, these results imply that teachers' instructional visions are related to some, but not all, aspects of their current instructional practice.

As a reminder, a "vision" represents a conception of the instructional practice that teachers *aspire* to enact.[53] Therefore, we expected that teachers' instructional visions might be more aligned with their *future* instructional practice than their current instructional practice. We found that this was, in fact, the case. Teachers' instructional visions at the outset of our study were significantly related to growth in their instruction during the first four years.[54] For example, on average, teachers who initially taught in conventional ways but articulated relatively sophisticated instructional visions improved their practice more than teachers who articulated less sophisticated visions. This finding suggests it is very important that teachers with conventional teaching practices are provided ample opportunities to value and develop a vision of ambitious and equitable teaching—and for those who are tasked with supporting teachers' learning to expect teachers' current practice and their vision to differ for a time. Developing such a vision likely requires opportunities to see it in action, and to develop an understanding of why the associated forms of teaching practice are critical for students' mathematical learning.

Views of Students' Mathematical Capabilities

Given the districts' commitments to ensuring that students from historically disadvantaged populations would have access to high-quality instruction and develop rigorous understandings of mathematics and productive mathematical dispositions, we found it especially important to investigate relations between teachers' views of their students and their instruction. Analyses across a large sample of teachers, in which we controlled for teachers' MKT and instructional visions, revealed that teachers' views of students were related to several aspects of instruction. First, teachers who described making adjustments to enable students who were currently facing difficulty to participate in rigorous mathematical activity tended to maintain the cognitive demand of high level tasks. On the other hand, teachers who doubted their students' capabilities tended to decrease the cognitive demand of high level tasks.[55] As described earlier in this chapter, maintaining the cognitive demand of challenging tasks is difficult. These findings suggest how important it is for professional development to explicitly focus on supporting teachers to learn how to adjust their instructional practice to support students who face difficulty, while maintaining rigorous goals for their learning.

Second, teachers' explanations of sources of students' difficulty in mathematics were significantly related to students' participation in high-quality mathematical discussions.[56] That is, students were, on average, more likely to have opportunities to participate in discussions in which they provided reasoning for their solutions if their teacher explained student difficulty in terms of instruction, as opposed to in terms of student deficits. Further, this relation was strongest in classrooms composed almost entirely of students of color. This means that, especially for teachers working primarily with students of color, it is important to attend to how teachers explain the sources of students' difficulty.

Further evidence of the importance of teachers' views of students comes from Sharpe's comparative study of MIST teachers who improved their practice versus those who did not.[57] She found that teachers who developed sophisticated visions and were provided with ongoing support to develop an ambitious and equitable vision of instruction did not develop the intended forms of practice unless they viewed their students from historically marginalized populations as capable of engaging in rigorous problem-solving. In other words, without productive views of students' mathematical capabilities, teachers are unlikely to develop ambitious and equitable instructional practices.

KEY TAKEAWAYS FOR INSTRUCTIONAL LEADERS

Here, we highlight key takeaways of the findings described above for instructional leaders (e.g., district leaders, school leaders, coaches) who are charged with designing and implementing strategies for supporting teachers' development of ambitious and equitable instructional practices.

While targeting teachers' mathematical knowledge is clearly important, doing so on its own will not result in large-scale instructional improvement. It was very common in both interviews and in our annual feedback and recommendations sessions for instructional leaders to suggest that when teachers did not enact ambitious and equitable forms of teaching, it was due to their insufficient "content knowledge." While content knowledge undoubtedly matters, as discussed above, teachers' knowledge of mathematics needs to extend beyond a personal understanding of key mathematical ideas; teachers need what Ball, Hill, and colleagues have called specialized content knowledge—that is the mathematical knowledge that is specific to the profession of teaching.[58] It includes how students' mathematical ideas develop over time, how to represent key mathematical ideas in accessible and accurate ways, and so forth.

However, even targeting sophisticated mathematical knowledge for teaching does not guarantee that teachers will enact ambitious and equitable teaching practices. In addition to MKT, teachers' instructional visions—that is, what they see as important to accomplish in the classroom—and teachers' views of their students' mathematical capabilities shape the extent to which they develop ambitious and equitable teaching practices.

In fact, teachers' instructional visions appear to be a leading indicator of improvement in classroom practice.[59] Teachers often develop facility with describing a more sophisticated vision before they enact it in their classroom. An implication of this for instructional leaders is to ensure that teachers have ample opportunities to develop a vision of high-quality mathematics instruction—and specifically to learn to notice and name important aspects of ambitious and equitable teaching.[60] As elaborated in chapter 4, to do so requires repeated opportunities to observe and discuss practice with others who are already accomplished in noticing key aspects of teaching ambitiously and equitably.

However, having sophisticated mathematical knowledge for teaching and articulating an ambitious and equitable vision of teaching does not guarantee teachers will enact desired practices in their classroom. Equally important are teachers' views of their students' mathematical capabilities. Recall the teachers discussed above who articulated a sophisticated vision of teaching, but did not enact such forms of teaching in their classrooms because they did not view their students from historically marginalized populations as capable of engaging in rigorous problem-solving.[61] Teachers' views of their students shape the instructional decisions they make in their classrooms, including the challenge of the tasks they pose to students, the extent to which they maintain or decrease the challenge of the tasks over the course of the lesson, and the extent to which they elicit and build on a wide range of students' thinking in whole-class discussions.

Supporting teachers to develop more productive views of their students' mathematical capabilities—especially of their historically marginalized groups of students—will require deliberate support in teachers' classrooms aimed at providing evidence that specific students, who teachers may have previously viewed as incapable, can, indeed participate in the desired forms of activity.[62] As an example, in one of our partner districts, an accomplished coach modeled high-quality classroom discussion in a classroom serving large numbers of English learners while the teacher observed. This supported the teacher in coming to view English learners as not only capable of engaging in math talk, but to also appreciate how doing so enabled English learners to develop deeper understandings of important mathematical ideas. As the teacher's instructional vision and views of students improved, he made progress in developing ambitious and equitable forms of practice.

Together, these findings suggest that professional learning designs need to target teachers' mathematical knowledge for teaching, instructional visions, and their views of their students' mathematical capabilities, in service of improving teachers' current instructional practices. We found that in most of our districts, professional learning for teachers primarily focused on teachers' mathematical knowledge, teachers' use of curricular materials, and to some extent teachers' visions. Clearly, broadening the focus of professional learning to the knowledge and perspectives described above is more difficult than, for example, only attending to teachers' content knowledge or understanding of the curriculum series. However, our findings question the extent to which most teachers will improve their instructional practices absent attention to these various forms of knowledge and perspectives.

Further, our findings suggest the importance of focusing professional learning on supporting teachers to learn how to maintain the challenge of cognitively demanding tasks. This includes learning how to support students who are currently facing difficulty to participate substantially in lessons organized around challenging tasks. As noted above, although most teachers in our partner districts introduced cognitively demanding tasks, they often lowered the challenge of tasks when they implemented them with their students, usually by suggesting a procedure to solve the task. Moreover, based on our findings, we have good reason to believe this happened more often in classrooms that served larger numbers of students from historically disadvantaged backgrounds.[63] Therefore, absent specific support in ensuring all students are enabled to support to engage in rigorous mathematical activity, we anticipate that teachers will continue to implement tasks in ways that reinforce long-standing inequities. We imagine that tackling this issue would entail supporting teachers to simultaneously develop mathematical knowledge for teaching, more sophisticated instructional visions, and more productive views of their students' mathematical capabilities—and thus could serve as a useful focus around which to organize professional learning.

FUTURE RESEARCH

In closing, we identify three areas for future research regarding relations between mathematics teachers' knowledge, perspectives, and practice in the context of ambitious instructional improvement initiatives. Developing ways to learn about teachers' current knowledge, perspectives, and instructional practice, as well as the ways in which teachers who vary in terms of their knowledge, perspectives, and current practice develop ambitious and equitable practices, is crucial in both designing for and assessing professional learning opportunities aimed at improving instruction at scale.

First, there is a need to investigate more deeply the relations between aspects of teachers' evolving knowledge, perspectives, and practice. For example, how do more sophisticated visions of high-quality mathematics instruction aid in instructional improvement? Why is it that teachers' instructional visions appear to be related to the challenge of the tasks they implement over the course of a lesson, yet not necessarily related to the quality of discussions they orchestrate in the classroom? Are there particular thresholds in terms of teachers' MKT, instructional vision, and/or views of their students that need to be attained to develop and sustain especially sophisticated forms of practice? And, how might these thresholds vary depending on aspects of the classroom, school, and district contexts in which teachers work? Just as importantly, how might teachers' knowledge, perspective, and practice develop in relation to specific professional learning supports? As we elaborate in chapter 5, given the challenges that districts face in terms of teachers' views of their students' capabilities, and the general lack of attention we documented to these views in our partner districts, it is especially important to both develop and investigate professional learning designs that are intended to support teachers to develop more productive views of their students' mathematical capabilities.

Second, the field of educational research is in need of relatively quick ways to assess teachers' instructional visions, their views of their students' mathematical capabilities, and their instructional practice. The existing measures that we used (MKT, teachers' classroom practice) and the two measures that we developed (instructional visions, views of students' mathematical capabilities) are relatively demanding to administer. Administering the measures and analyzing the resulting data required a significant amount of time and effort on the part of researchers; as such, we tended to provide annual feedback regarding teachers' knowledge, perspectives, and practice. However, instructional leaders need much more frequent assessments of teachers' knowledge, perspectives, and practice in order to tailor the design of their professional learning supports to the strengths and needs of teachers on a regular basis.

We have come to see great value in what the Carnegie Foundation for the Advancement of Teaching calls "practical measures," which are designed to provide practitioners with frequent, rapid feedback that enables them to assess and improve their practices.[64] As an example, we recently developed three-minute student surveys to assess the quality of small-group and whole-class discussions.[65] The resulting data on the quality of instruction is proving useful in informing teachers' efforts to improve their practice, as well as coaches' and math leaders' work with teachers. We have found that similar to developing traditional research measures, it is challenging to develop practical measures that provide reliable data and are easy to administer and analyze. However, we see great potential in developing such measures, especially in terms of the frequency with which assessments can be made, and therefore the ways in which teachers can adjust aspects of their teaching, and instructional leaders can adjust the design of professional learning supports.

Third, there is a need for assessments that target additional aspects of teachers' knowledge, perspectives, and practice central to enacting an ambitious and equitable vision of teaching. One example concerns teachers' knowledge of how students develop core mathematical ideas, like proportional reasoning, over time.[66] Studies of elementary mathematics teaching have pointed to the importance of teachers developing such understandings.

For example, studies of teachers who participated in the Cognitively Guided Instruction program illustrate the value of teachers making sense of a trajectory of the strategies young children tend to develop to solve problems involving operations on whole numbers.[67] This foundation enabled teachers to make deliberate instructional decisions in response to their students' current understandings. Having assessments of how students develop core mathematical ideas over time could support instructional leaders to more efficiently establish appropriate learning goals for teachers. In addition, we anticipate that teachers' understanding of a trajectory of student thinking may also support their development of productive views of their students' mathematical capabilities.[68] Namely, a trajectory suggests that students have resources on which to build, and can and will learn mathematics with understanding when engaged in appropriate, challenging tasks.

Another example concerns ways to assess the extent to which instruction is equitable. Having such assessments are important for making visible aspects of instruction to target in professional learning. In both research and practice, we tend to target that which is assessed. At present, there are few tools for assessing concrete aspects of instruction that appear important for advancing equity in the classroom; not coincidentally, there is limited attention to equitable instruction in both the literature on accomplishing instructional improvement at scale, and, as we found in our districts, in professional learning settings. As we noted above, members of the MIST research team have created measures for assessing the presence and quality of specific aspects of equitable instruction that either literature suggested or that emerged through study of MIST video-recordings of classroom instruction.[69] Another example is the work of Daniel Reinholz and Niral Shah, who have developed an observational tool known as EQUIP: Equity Quantified In Participation.[70] EQUIP targets how opportunities to participate in whole-class discussions are "distributed across socially constructed markers of difference (e.g., gender, race)."[71] Whereas the examples above are traditional research measures, for the reasons described above, there is also a need to develop equity-specific practical measures.

As equity-specific measures are more widely validated, it is important to investigate relations between teachers' development of equitable teaching practices and other aspects of their experience, knowledge, and perspectives and their participation in professional learning. Such investigations would allow researchers and practitioners to better understand how to design professional learning that results in ambitious *and* equitable teaching.

CONCLUSION

In this chapter, we first described a vision of ambitious and equitable mathematics instruction, which constitutes the core of the coherent instructional system. We then shared our findings regarding how teachers' knowledge and perspectives shapes the extent to which they develop the forms of practice specified in this vision. The vision, as well as our findings regarding teachers' knowledge, perspective, and practice, is intended to orient decisions about what goals to pursue for both student and teacher learning, what curricular materials to invest in, the assessments that are best suited for tracking and informing improvement, the nature of additional supports for students who currently struggle, and the focus of professional learning initiatives. Subsequent chapters will engage in discussion of each of these issues in turn, and will put forth a set of takeaways regarding how to design and implement strategies that are likely to result in the improvement of teaching and learning.

FOUR

Overview of the Teacher Learning Subsystem

KARA JACKSON, ILANA HORN, AND PAUL COBB

THE VISION OF AMBITIOUS and equitable instruction, as discussed in chapter 3, represents a set of goals for teachers' learning. That is, to enable students to attain rigorous goals for mathematics learning teachers must, for example, learn to choose and introduce rigorous tasks, elicit and build on student thinking in the moment, support students to make sense of one another's thinking, and enable students to make connections between multiple representations of mathematical ideas. Moreover, teachers must ensure that each and every student is being encouraged and supported to participate substantially in all aspects of classroom activity.

Planning for and enacting ambitious and equitable mathematics teaching is a major shift from what happens in most mathematics classrooms in most schools and districts.[1] As illustrated in chapter 3, this was true for classrooms in our four partner districts as well. For most teachers participating in the MIST project, the vision of ambitious and equitable instruction that oriented their districts' instructional improvement efforts was a significant departure from their current teaching practice.[2]

Decades of research suggest that learning to develop the intended forms of practice is by no means trivial.[3] As discussed in chapter 3, shifting from conventional mathematics teaching to ambitious and equitable mathematics teaching requires that teachers reimagine core aspects of their practice and the relations between them: what it means to be a teacher, what mathematics is, and how students learn mathematics. It requires that teachers develop a distinct vision of what counts as high-quality mathematics instruction, deepen their mathematical knowledge for teaching, reconsider their current views of their students' mathematical capabilities, and develop new forms of teaching practice. In light of these learning demands, the majority of mathematics teachers in our partner districts required extensive support if they were to develop ambitious and equitable teaching practices.

Our partner districts recognized this need and, like many districts, designed and implemented various forms of support for teachers' learning. However, as we will clarify below, we often found that these supports were not coordinated and that they typically differed in their intent and focus. Over the course of MIST, it proved essential that districts conceptualize teacher learning as a *system* in which various supports are *connected and coordinated*. It is for this reason that a *teacher learning subsystem* is a critical element of the coherent instructional system in the MIST theory of action. (See figure 1.1 in chapter 1.)

In this chapter, we first provide the reader with a description of the key types of support for teachers' learning that constitute important facets of a teacher learning subsystem: pull-out professional development (PD); teacher collaborative time; content-focused instructional coaching; and teacher advice networks. We then provide an illustration of a potentially productive teacher learning subsystem that was designed and enacted in one of our partner districts. Against this, we identify key design principles of a teacher learning subsystem that has the potential to support teachers' development of ambitious and equitable instructional practices. We also highlight the need to coordinate the focus of professional learning for other key role groups (e.g., school leaders, coaches) with the focus of professional learning for teachers.

KEY TYPES OF SUPPORT FOR TEACHER LEARNING

We found that our partner districts tended to design and implement a similar set of supports for teachers' learning, which are common in districts pursuing instructional reform. The most ubiquitous type of designed support was *pull-out professional development (PD)*. As discussed in chapter 5, pull-out PD refers to professional development in which teachers leave their school campuses to engage in professional learning, usually (but not necessarily) by grade level. It often involves teachers from multiple schools, and is typically designed and led by either a district's mathematics leaders or by external contractors. Another common form of support is what we refer to as *teacher collaborative time*, which we discuss in chapter 6. Although the design and implementation of teacher collaborative time varied by district and school, in general, it involved providing mathematics teachers, usually grouped by grade level, with regularly scheduled time to collaborate on improving teaching. Most of our districts also designed and implemented some form of *content-focused instructional coaching*, including district-based and school-based coaching. In chapter 7, we clarify that content-focused coaching offers the opportunity for teachers to work one-on-one in their classrooms with a relative expert. In the chapters that follow, we discuss the potential of and challenges associated with each of these three designed forms of support, respectively, as key facets of a teacher learning subsystem. In chapter 8, we discuss a fourth type of support—*teachers' advice networks*—that we found to be central in supporting teachers' learning. Teachers' advice networks emerge as teachers informally seek one another out for support in teaching. As such, they differ from the other three types of support in that a district or school cannot *design* an advice network per se. However, as we clarify in chapter 8, MIST findings indicated that the quality of *designed* supports (i.e., pull-out PD, teacher collaborative time, content-focused coaching) influences the conditions in which and the quality of networks that emerge. We therefore treat teachers' advice networks as a central facet of a teacher learning subsystem.

ILLUSTRATION: PRODUCTIVE SYSTEM OF DESIGNED SUPPORTS FOR TEACHER LEARNING

To illustrate what we mean by a teacher learning subsystem, we describe how central office leaders in one of our partner districts organized teacher learning around Formative Assessment Lessons (FALs) in six middle schools in the final year of the MIST study. FALs are organized around cognitively demanding mathematics tasks and require students to engage in disciplinary reasoning. They were designed by the Mathematics Assessment Resource Service to support teachers to develop formative assessment practices, whereby teachers deliberately inquire into student thinking and use what they learn to inform their instructional decisions. Central office leaders viewed organizing professional learning around the FALs as a way to both increase the rigor of mathematics in classrooms and to support teachers to develop high-leverage practices associated with ambitious and equitable instruction (e.g., pose cognitively demanding tasks, elicit and build on student thinking to pursue an instructional agenda).

Central office leaders coordinated district-wide, pull-out PD and school-based teacher collaborative time around the implementation of FALs. Eight full day pull-out PD sessions were scheduled throughout the year. During these sessions, school-based teams of teachers, a school-based coach, and the principal engaged in upcoming FALs lessons. Teams were introduced to central formative assessment practices, worked through the mathematical tasks, identified important mathematical ideas, considered how students' development of these ideas would be supported in the lessons, and mapped specific lessons to state standards for mathematics. However, district leaders recognized that such sessions would not, by themselves, be sufficient to support teachers in becoming proficient at eliciting and responding to student thinking in the moment. Therefore, in addition to pull-out PD, central office leaders expected the school-based teams to meet at least biweekly to further refine their plans for upcoming lessons and to analyze students' work after each FAL. In addition, a coach, accomplished in enacting the intended forms of instruction and in supporting teachers' learning, provided one-on-one coaching for teachers as they implemented FALs in their classrooms. Early in the year, the coach co-planned and co-taught lessons with teachers, but later in the year she co-taught less and instead observed instruction and provided targeted feedback.

The approach that this district took regarding implementing the FALs initiative contrasts sharply with what happens in most districts and schools. Most often—and this was certainly the case in most of the MIST districts—one-on-one coaching does not directly build on either the district-wide PD in which the teachers have participated or their work during teacher collaborative meetings. Additionally, what happens in teacher collaborative time remains disconnected from what happens in pull-out PD. This makes for an incoherent set of supports that implicitly communicates to teachers that they should select the practices that best suit them, thereby undermining the potential of any one form of support to have any lasting impact.

KEY DESIGN PRINCIPLES OF A POTENTIALLY PRODUCTIVE SYSTEM OF SUPPORTS FOR TEACHER LEARNING

In what follows, based on extant research as well as MIST analyses, we suggest what district leaders, school leaders, and facilitators of professional learning should attend to when designing a productive system of supports for teachers to develop ambitious and equitable teaching, like the one described above. (See table 4.1 for a summary of the design principles.) The chapters that follow which focus on designed supports (i.e., pull-out PD, teacher collaborative time, content-focused coaching) refer back to these principles. Of course, the form that the designed supports take will necessarily vary, depending on local resources and constraints. For example, as will be elaborated in chapter 7, in our partner districts, decisions about a district's coaching model depended on local conditions, including available expertise in the district, funding, number of schools in

TABLE 4.1 Design principles for a potentially productive teacher learning subsystem

DESIGN PRINCIPLES FOR A POTENTIALLY PRODUCTIVE TEACHER LEARNING SUBSYSTEM
1. The focus of various supports for teachers' learning (e.g., district-wide pull-out professional development, teacher collaborative time, content-focused coaching) should be coherent and tightly connected so that the goals for improving classroom practice being worked on in one type of support are built on and elaborated in other types of support.
2. Teachers should have opportunities to work on improving their classroom practice with the same colleagues over time.
3. The activities in which teachers engage should be close to instructional practice and organized around high-leverage aspects of teaching practice.
4. Activities should include pedagogies of investigation and enactment, organized around the development of specific forms of practice.
5. Facilitators of designed supports for teacher learning should have expertise in ambitious and equitable teaching, as well as practical knowledge of teacher development.

a district, and so forth. Similarly, system resources and constraints need to be taken into consideration when deciding if, and if so, what to focus on in district-wide PD and/or whether to invest in regular time for teachers to collaborate.

While the design and implementation of the facets of a teacher learning subsystem will necessarily vary by district and school context, as a first principle, it is essential that district leaders, school leaders, and facilitators are intentional in ensuring that the types of designed support are deliberately coordinated as a *system*. By this, we mean that the focus of professional learning in the various types of designed supports should be coherent and tightly connected, so that the aspects of classroom practice being worked on in one type of support are connected and elaborated on in another. For example, note that in the illustration above, the pull-out PD for the FALs laid a foundation for teaching lessons that center on eliciting and building on student thinking to achieve an instructional agenda. This foundational work was then connected to and elaborated on during both school-based teacher collaborative time and one-on-one coaching. Furthermore, the coach's one-on-one work with teachers could then inform the focus of both school-based teacher collaborative time and district-wide pull-out PD sessions.

As a second principle, for teacher professional learning to support teachers to significantly reorganize their practice, it needs to be sustained over time.[4] This principle questions the value of "one off" PD sessions, which we observed in each of our districts, in which teachers are introduced to an isolated aspect of teaching absent follow-up support. For example, in an earlier year in the MIST project, the district described above provided a one-off, isolated session focused on formative assessment practices at the beginning of the school year that our analysis suggested had limited, if any, impact on teachers' practice. We found that, at best, the one-off session resulted in only minimal changes to teachers' practice (e.g., as a result of the session, some teachers administered exit tickets to elicit what their students had learned in a lesson). The one-off session did not, however, support teachers to elicit students' thinking in the moment of instruction, let alone know how to respond to students' thinking to advance an instructional agenda. This contrasted sharply with the outcomes of the FALs initiative described above, in which teachers had sustained opportunities to improve how they elicited and responded to student thinking.

Moreover, it is important that the same group of teachers have opportunities to work together to improve their classroom practice for an extended time.[5] This might happen in the context of pull-out PD and in teacher collaborative time. Collaborating together on improving teaching runs counter to the norms of the teaching profession, in which teachers are typically isolated from their colleagues and norms of privacy and autonomy persist.[6] Sustained collaboration can both foster the trust necessary to make classroom practice public and open to scrutiny, and provide a set of shared experiences to support teachers' ongoing sensemaking. In the illustration above, it was important that school-based teams attended the pull-out PD on the FALs and had regular time to work together back on campus, and thus develop a culture of collaboration. It is likely that this impacted the development of supportive advice networks.

As a third design principle, designed supports for teachers' learning should be close to practice by focusing on issues central to instruction and by being organized around the materials that teachers use in their classroom.[7] For example, as discussed in chapter 9, it is important to organize supports for teachers' learning around the instructional materials and associated tools (e.g., curriculum frameworks) that they are expected to use on a daily basis. In the illustration above, the supports were organized around specific lessons that teachers were expected to implement in their classroom.

In addition, research has indicated the value in focusing supports for teachers' learning on aspects of practice that are likely to make a critical difference in students' learning opportunities—what are often called *high-leverage practices*.[8] In the illustration above, supports for teachers' learning focused on the implementation of cognitively demanding tasks, which, as described in chapter 3, matters crucially for the kind of mathematical understandings and reasoning that students develop.[9] Orchestrating whole-class discussions in which students explain and justify their reasoning is another example of a high-leverage practice, around which our partner districts organized professional learning.[10] As described in chapter 3, most teachers conducted show-and-tell discussions that do not support students in learning from others' explanations.[11] Improving the quality of whole-class discussions was essential to supporting broader student learning and participation. Moreover, focusing on this practice can also support teachers to anticipate student thinking on specific tasks, consider which student solutions to discuss and in what order, press on students' ideas, and develop their mathematical knowledge for teaching—all of which are necessary aspects of developing ambitious and equitable teaching practices.[12]

As a fourth design principle, studies of professional learning in which the intended forms of practice represent a significant shift for learners suggest that it is essential that a system of designed supports involve coordinated pedagogies of investigation and of enactment.[13] *Pedagogies of investigation* are typical in PD settings and entail analyzing and critiquing records of practice, such as video-recordings of teaching or student work samples.[14] These activities support teachers' learning by providing images of high-quality practice and/or opportunities to reflect on current instruction. However, *only* engaging in pedagogies of investigation does not support teachers to enact complex and novel forms of practice.

In addition, teachers need opportunities to actually try out intended forms of practice with targeted feedback from an expert, in settings of somewhat reduced complexity. Pam Grossman and colleagues have referred to such opportunities as *pedagogies of enactment*.[15] For example, in a common pedagogy of enactment that occurred in the MIST districts, teachers would co-plan an upcoming lesson with a coach and then try out targeted forms of practice in the context of the lesson, with in-the-moment feedback from the coach. Another example of a less common pedagogy of enactment involved rehearsing for upcoming instruction, usually in the context of teacher collaborative time.[16] For example, a teacher might try out how she intended to introduce a task, with her colleagues acting as students, with the goal of identifying problematic aspects of the introduction; this allowed the rehearsing teacher to adjust her lesson plan prior to actually teaching the lesson.

As illustrated above, from the perspective of a teacher learning subsystem, pedagogies of investigation should be coordinated with pedagogies of enactment. For example, in the FALs initiative, teachers primarily engaged in relevant pedagogies of investigation in district-wide pull-out PD (e.g., they analyzed upcoming tasks and lessons). They then had opportunities to engage in related pedagogies of investigation and enactment at their schools. Teachers tried out specific formative assessment practices in the context of a FAL that they had co-planned with the support of an expert coach (pedagogy of enactment). These enactments were then followed up with investigations, such as a group analysis of the resulting student work, in teacher collaborative time.

A fifth, overarching design principle of a system of designed supports for teachers' learning focuses on the importance of facilitators' expertise. For any designed support to be effective, it is crucial that the support includes co-participation with accomplished others.[17] Without someone with substantial expertise to facilitate teacher learning, even the most well-intentioned activity will fall short. Indeed, a recurrent difficulty in some of our partner districts was that those charged with supporting teacher learning had not yet developed the intended forms of teaching practice themselves. This undermined the potential of the activity to support teachers' development of the intended forms of practice. Understandably, teachers also failed to see the facilitators as experts from whom they could learn.

That said, facilitators' expertise as teachers was necessary but not sufficient. We also found that facilitators who were accomplished mathematics teachers were not necessarily prepared to support others' development of ambitious and equitable practices. As we will elaborate on in the chapters that follow, in addition to being accomplished teachers, MIST analyses highlighted the importance of facilitators having a working knowledge of teacher development. For instance, MIST analyses indicated that effective facilitators assess teachers' current practices in relation to a set of goals for teachers' learning.[18] They then use those assessments to identify short-term goals for teachers' learning and to select and design activities to move teachers towards those goals.[19] For example, in the FALs illustration above, facilitators of pull-out PD worked closely with the coach who provided teachers with one-on-one support to determine what made sense to focus on in each pull-out PD session, in light of the teachers' current practices. Similarly, the coach adjusted her individual work with teachers as well as her work with school-based teams of teachers, depending on what happened in the lessons teachers piloted in their classrooms. In addition, MIST analyses and extant research indicates that effective facilitators attend to teachers' motivations to engage in professional learning.[20] Effective facilitators build relationships with teachers, and help teachers find a purpose in engaging in professional learning activities. Having met the teachers on their terms, effective facilitators are then in a position to support teachers to develop new ways to interpret and respond to challenges that teachers face.[21]

COORDINATING PROFESSIONAL LEARNING ACROSS ROLE GROUPS

In chapters 5–8, we discuss each of the key types of support for teachers' learning that constitute important facets of a teacher learning subsystem: pull-out PD; teacher collaborative time; content-focused instructional coaching; and teacher advice networks. Throughout, we will emphasize the importance of coordinating the focus of the system of supports for teachers. In addition, in the course of our partnerships with the four districts, we found that it is essential to coordinate the system of supports for teachers' learning with supports for other role groups' learning, especially school leaders and coaches, as well as central office leaders.[22] An emphasis on coordination of professional learning across role groups is consistent with the *systems perspective* we described in chapter 1. A systems perspective on instructional improvement requires attention to the conditions that shape the implementation of supports for teachers' learning. As we will highlight especially in chapters 7, 12, and 13, the expertise of coaches, school leaders, and central office leaders impacts the nature of the expectations communicated to teachers regarding what they should be working towards, instructionally, and the nature of the supports teachers are provided. And, as will be substantiated in the forthcoming chapters, our findings suggest that most coaches, school leaders, and central office leaders needed sustained professional learning opportunities to develop the capabilities necessary to press and support teachers to, in turn, develop ambitious and equitable teaching practices.

As an example of how professional learning might be coordinated across role groups, we focus on the attention given to principals' professional learning in the FALs initiative we described above. Principals were deeply involved in this work. As noted earlier, principals co-participated in the pull-out PD sessions for teachers. In addition, they were provided with administrative coaches, who supported them to implement what was focused on in the pull-out PD on their campuses. Moreover, principals were expected to observe teachers' implementing FALs in their classroom with their administrative coach and with the mathematics coach that was supporting the teachers. Through joint observations, principals were supported to deepen their understanding of the features and importance of cognitively demanding tasks, and of the importance of building instruction based on students' thinking. In addition, the principal and the coach jointly provided feedback to teachers. This allowed principals to deepen their understanding of how to provide feedback that communicates appropriate instructional expectations. More generally, the deliberate coordination of professional learning for principals and teachers supported the principals' development of an instructional vision consistent with that of the teachers and coach, and thus of what should be happening in classrooms. It is this kind of coordination that we suspect is necessary, if principals are to implement a coherent system of supports for teachers in their schools.

FIVE

Pull-Out Professional Development for Teachers

KARA JACKSON, MEGAN WEBSTER, AND JONEE WILSON

IN THIS CHAPTER, we focus on a common form of support for teacher learning, pull-out professional development (PD).[1] Pull-out PD refers to PD in which teachers from multiple schools engage in professional learning, usually, but not necessarily, by grade level. It is typically designed and led by either a district's mathematics leaders or by external contractors. In what follows, we discuss the potential of pull-out PD to support teacher learning, even as we argue that high-quality pull-out PD is, by itself, frequently insufficient to support teachers in improving the quality of their classroom instruction. We then make recommendations specific to designing and facilitating pull-out PD as *one* facet of a subsystem of supports for teachers' learning. These recommendations are informed by both the research literature and our findings, which are based on the annual feedback and recommendation cycles as well as retrospective analyses. Last, we identify key areas for future research regarding pull-out PD, specific to instructional improvement at scale.

THE POTENTIAL AND LIMITATIONS OF PULL-OUT PROFESSIONAL DEVELOPMENT AS A SUPPORT FOR TEACHERS' DEVELOPMENT OF AMBITIOUS AND EQUITABLE PRACTICES

Pull-out PD is the primary form of support for teachers' learning in many districts. The leaders of our partner districts regarded pull-out PD as a means of communicating a consistent message to teachers regarding expectations for mathematics learning and teaching. In addition, pull-out PD promoted cross-campus collaboration in our partner districts. As such, it often provided teachers who worked in schools with limited expertise in ambitious and equitable teaching with opportunities to collaborate with colleagues with expertise from other schools and from the district central office, and with expert outside consultants. Such cross-campus collaboration was especially important for teachers who taught in small middle schools and therefore did not have a mathematics teacher colleague on their campus with whom to collaborate.

That said, while pull-out PD may play a significant role in supporting district-wide instructional improvement, it is usually insufficient to support instructional improvement at scale. Why is that the case? As we elaborated in chapter 4, teachers are unlikely to develop ambitious and equitable practices unless they have opportunities to work with more expert others on intended forms of practices on a regular basis, specific to the resources and challenges associated with their own schools and classrooms. That is, unless teachers attend pull-out PD together (e.g., as a grade-level team) *and* continue to meet regularly on campus to work on instructional issues, they are frequently "on their own" to determine whether and how to implement what they learned in pull-out PD. Working alone rarely results in the significant shifts in practice entailed in enacting a vision of ambitious and equitable teaching.

Clearly, the value of pull-out PD depends crucially on its focus and the quality of its design and facilitation.[2] We draw on the design principles for a potentially productive teacher learning subsystem laid out in chapter 4 to characterize high-quality pull-out PD (see table 4.1). In doing so, we note when pull-out PD, on its own, might only partially meet a particular criterion. As a first principle PD must be tightly connected with other forms of support that are more proximal to the classroom, such as one-on-one coaching or teacher collaborative time. That is, the learning goals of a sequence of pull-out PD sessions should be both connected to and further elaborated on in the context of other types of support for teachers' learning. In chapter 4, we illustrated how one of our partner districts productively coordinated a sequence of pull-out PD sessions with school-based teacher collaborative time and one-on-one coaching focused on improving the rigor of tasks and teachers' formative assessment practices.

As a second principle, teachers must engage in sustained PD with the same colleagues over time.[3] It is well established that one-off PD sessions do not support teachers to substantially improve their current practices.[4] If pull-out PD is to support teachers' development of ambitious and equitable forms of practice, it must be designed as a sequence of sessions in which goals for teachers' learning shift over time as their current practices develop. For example, in pull-out PD led by Sabrina, a highly regarded and successful PD leader in one of the MIST districts, a series of monthly sessions progressed from an initial focus on identifying the cognitive demand of curricular tasks, to anticipating student thinking specific to upcoming tasks, to developing routines regarding identifying student solutions that might make for a useful whole-class discussion and purposefully sequencing a discussion of such solutions, to then considering key practices of introducing challenging tasks to ensure all students could engage with important mathematics.[5] Sabrina purposefully sequenced the focus of the sessions with respect to her assessment of teachers' developing practices and a sense of what practices might be more difficult to enact than others and which practices depended on having established prior ones. In addition to purposefully sequencing pull-out PD sessions, Sabrina organized the teachers to participate in the pull-out PD sessions in consistent, grade-level cohorts. It is important for the same group of teachers to work together across the sessions (e.g., by school, by grade level) so that they can, hopefully, develop the trust necessary to open up their practice to scrutiny.[6]

There can be value in one-off PD sessions when the goal is to support teachers to incorporate *minor* extensions or elaborations to their current practice. For example, we found in one of our partner districts, there was value in a one-off pull-out PD session for all middle-grades mathematics teachers in which a new pacing guide was introduced. The pacing guide provided information regarding how to sequence particular lessons, and when to teach which lessons. In other words, using the pacing guide did not require teachers to make changes to *how* they taught – which would have arguably required substantial reorganization of most teachers' current practices and therefore sustained PD. Rather, it only required them to change *when* they taught their material. It was therefore not surprising that the majority of teachers reported using the pacing guide in an interview several months after the session.

As a third principle, the focus of the pull-out PD should be close to practice and should be organized around high-leverage teaching practices.[7] As Deborah Ball and David Cohen argue, "To learn anything relevant to performance, professionals need experience with the tasks and ways of thinking that are fundamental to the practice."[8] In the example of the high-quality pull-out PD which Sabrina led, choosing cognitively demanding tasks, planning and orchestrating whole class discussions that support conceptual learning, and introducing tasks to support all students' substantial engagement while maintaining the cognitive challenge all represent fundamental tasks of teaching ambitiously and equitably. These are the sorts of high-leverage practices on which pull-out (and other forms of) PD should focus. It is important that professional learning focused on developing such high-leverage teaching practices are grounded in teachers' classroom instructional materials (e.g., textbooks, curriculum frameworks). As discussed in chapter 9, ongoing PD is required if teachers are to learn how to use materials that suggest significant changes in practice.[9]

The fourth principle concerns the importance of including both pedagogies of investigation and enactment, organized around the development of high-leverage forms of practice, in professional learning settings.[10] In our view, pull-out PD is particularly well-suited for pedagogies of investigation because large numbers of teachers from a variety of schools are typically involved. For example, in the example discussed above, pull-out PD was well-suited for supporting teachers across schools to develop common frameworks for analyzing the cognitive demand of tasks, student work, and video-recordings of high-quality instruction. This not only contributed to supporting individual teachers' practices; it also appeared to support the development of a district-wide vision of what counted as high-quality mathematics teaching and learning.

The value in pedagogies of enactment concerns opportunities to try out intended forms of practice in settings of reduced complexity with targeted feedback from an expert.[11] It is more difficult to engage in pedagogies of enactment that closely approximate classroom practice in a pull-out PD situation precisely because the teachers have been "pulled out" of classroom settings. It is certainly possible to engage in pedagogies of enactment in pull-out PD; in the example above, Sabrina had groups of teachers across campuses rehearse an upcoming whole class discussion with other teachers acting as students. However, in a pull-out setting, it is difficult to closely approximate the conditions in which the teacher would actually lead the discussion. For example, it is unlikely that the teachers acting as students were familiar with the range of ways in which the rehearsing teacher's students might solve a given task, and therefore the rehearsal might have felt inauthentic. Moreover, it is difficult for an expert facilitator to provide the needed scaffolding to multiple groups of teachers involved in simultaneous rehearsals. On the other hand, as will be elaborated in the chapters that follow, teacher collaborative time and instructional coaching are well-suited for both pedagogies of investigation and of enactment. Consequently, pull-out PD that only includes pedagogies of investigation can satisfy this fourth principle if it is connected to and elaborated on in teacher collaborative time and/or one-on-one coaching. For example, teachers might first analyze video-recordings of teachers introducing cognitively demanding tasks in district pull-out PD, and then enact introducing similar tasks with their colleagues in teacher collaborative meetings that are facilitated by a coach. This coordinated approach focuses on supporting teachers' development of a specific set of instructional practices over time, using materials that are central to their instruction.

Finally, if pull-out PD is to support teachers' development of ambitious and equitable forms of practice, it must be led by a skilled facilitator who has substantial expertise in both teaching and supporting teacher learning.[12] MIST analyses across pull-out PD, teacher collaborative time, and coaching all pointed to the crucial importance of the facilitator.[13] The extent to which each of the supports actually enabled teachers to improve the quality of their instruction appeared to depend critically on the expertise of the facilitator. Below, we report on both MIST-specific findings and the extant research regarding what facilitators of pull-out PD need to know and be able to do in order to support teachers to shift from conventional teaching to ambitious and equitable teaching.

DISTRICTS' CHALLENGES SPECIFIC TO PULL-OUT PD

All four of our partner districts provided significant amounts of professional learning, including pull-out PD that was designed and led primarily by district mathematics specialists or, in some cases, external contractors. Results from our annual teacher surveys show that, across the first four years of the MIST study, 39 percent of teachers reported receiving sixteen to thirty-five hours of professional learning specific to mathematics teaching (which included PD workshops or seminars and coaching) each year, and 47 percent reported attending more than thirty-five hours each year. This is considerably more than the amount of PD reported by teachers nationally. For example, on the 2012 National Survey of Science and Mathematics Education, 23 percent of middle-grades mathematics teachers reported attending a total of sixteen to thirty-five hours of PD specific to mathematics teaching (which included PD workshops or seminars and teacher collaborative time) in the past *three* years, and 31 percent of the teachers reported attending more than thirty-five hours of PD in the past *three* years.[14]

Although our partner districts offered a considerable amount of pull-out PD that focused on the development of ambitious and equitable teaching practices (e.g., how to use the intended instructional materials, how to develop specific high-leverage teaching practices), its quality varied considerably. Teachers' reports of the impact of PD on their practice reflected the uneven quality. There were certainly exceptions, but in many of the districts and across many years, teachers reported in interviews that mandatory pull-out PD did not impact their practice to any significant extent.[15] Teachers also indicated the PD lacked depth and relevance, and they were concerned that the PD was disconnected from ongoing, school-based professional learning opportunities such as teacher collaborative time and one-on-one coaching. These findings are not unusual; research suggests that most pull-out PD lacks focus and coherence.[16]

Our annual feedback and recommendation reports and retrospective analyses indicate that our partner districts encountered three specific challenges with respect to pull-out PD that impacted the extent to which the PD was likely to support teachers to develop ambitious and equitable practices. As indicated above and in chapter 4, the first challenge concerns the lack of coordination with other school-based forms of support for teacher learning, like teacher collaborative time or one-on-one coaching.

A second challenge concerns the lack of sustained attention to issues of equity in the majority of pull-out PD. In all of our partner districts, a significant amount of pull-out PD was focused on developing high-leverage teaching practices (e.g., choosing rigorous mathematical tasks, orchestrating productive whole-class discussions). However, teachers reported that they received little support in enacting these teaching practices with diverse learners. There were occasional one-off sessions that focused on issues of equity, for example, supporting English learners in mathematics or supporting African American students more generally. However, these sessions did not, by and large, meet the criteria of high-quality professional learning. It was therefore not surprising that in our interviews, teachers reported that it was difficult to know what to do, *in practice*, on the basis of the equity-specific pull-out PD sessions.

As indicated above, a third challenge our partner districts faced concerned the facilitation of pull-out PD. Although there were exceptions, by and large, our partner districts struggled to identify local mathematics leaders who had the knowledge and practice to support teachers' development of ambitious and equitable teaching practice.[17] Typically, the districts identified accomplished teachers who had developed sophisticated teaching practices to lead pull-out PD. However, as indicated in chapter 4, being an accomplished teacher is necessary but not sufficient for supporting other teachers' learning. When we analyzed video-recordings of pull-out PD, we found facilitators often chose to enact particular pedagogies that are promising for supporting teacher learning, such as co-planning instruction or analyzing video-recordings of instruction.[18] However, the activities as enacted were unlikely to result in improvement to teachers' current practice because of the quality of the facilitation. For example, in a PD session that focused on co-planning with colleagues for an upcoming unit of instruction, facilitators gave teachers only minimal direction before they began working. As a result, teachers approached co-planning as a "divide and conquer" activity in which they divvied up who would take responsibility for planning various lessons. Teachers did not solve any of the mathematical tasks in the unit, and facilitators did not press them to articulate goals for students' learning. The divide-and-conquer approach to co-planning reflected how teachers typically engaged in co-planning on their campuses. While it is efficient, dividing and conquering does not allow teachers to make sense of the key mathematical ideas of a unit, or to consider how students might develop those ideas. In addition, it does not provide opportunities for novice teachers to learn from more accomplished teachers where students might struggle and how to support the participation of a diverse group of students. In other words, it is unlikely that co-planning in the pull-out PD session actually supported teachers to analyze and improve their current planning practices because the facilitation was not oriented by clear goals for teachers' learning.

Additionally, we found across the districts that in PD sessions with multiple types of activities, the activities were rarely connected to one another. For example, in a 2.5 hour pull-out PD session focused on supporting teachers to engage in formative assessment, the facilitators a) led a discussion about the distinction between formative and summative assessments, b) modeled how to introduce a lesson productively, c) engaged teachers in analyzing student work samples and providing written feedback, d) asked teachers to generate examples of different assessment techniques and describe the pros and cons of each, and e) modeled how to use technology to quickly check for student understanding. Although each of these activities was potentially worthwhile, it was difficult to discern the facilitators' specific learning goals for teachers, and how these different activities could support teachers in achieving these goals.[19]

Furthermore, although facilitators elicited teachers' ideas at some point in most sessions, including those described above, they typically did not press teachers to elaborate their thinking or attempt to build on teachers' contributions in meaningful ways. For example, they rarely pressed teachers to elaborate the rationale for their contributions, checked to see if teachers understood others' contributions, or supported teachers to make connections between contributions.[20] That is, in most of the MIST districts, pull-out PD was characterized by a culture of politeness that, as Ball and Cohen describe, results in "reaffirm[ing] teachers' needs to do what fits them personally" and does not lead to developing the ambitious and equitable forms of practice that were at the heart of the districts' instructional improvement agendas.[21]

The challenge of facilitation is not unique to our partner districts; several scholars have identified the lack of local expertise in facilitating high-quality PD as a major impediment to achieving significant instructional improvement at scale.[22] Other research has identified similar challenges specific to novice facilitators, including difficulties in pressing on teachers' reasoning and in identifying goals for teachers' learning.[23] As we will elaborate below, the challenges led us to investigate designs for supporting the development of novice facilitator's practices in collaboration with our partner districts in the latter years of MIST.

TAKEAWAYS FOR THE DESIGN AND FACILITATION OF HIGH-QUALITY PULL-OUT PROFESSIONAL DEVELOPMENT

In light of these challenges, as well as the research literature and our findings, we highlight four key takeaways regarding the design and facilitation of pull-out PD that has the potential to support teachers' development of ambitious and equitable teaching practices.

First, ensure that the focus of pull-out PD is connected to and elaborated on in the context of school-based professional learning. As described above and in chapter 4, it is essential that pull-out PD is treated as just one facet of a coordinated, coherent subsystem of supports for teachers' learning. That is, the focus of pull-out PD sessions should be coordinated with the focus of teacher collaborative time and/or one-on-one coaching. Absent deliberate and scaffolded opportunities to practice new forms of teaching in the context of their classrooms, it is unlikely teachers will develop ambitious and equitable forms of teaching practice.

Second, ensure that attention to issues of equity is integrated into pull-out PD that aims to support teachers' development of high-leverage teaching practices.[24] As we argued in chapter 3, it is crucial that teachers view all their students as mathematically capable and that they know how, in practice, to engage all students, especially students who have been historically marginalized, in rigorous mathematical activity. As we discussed above, our partner districts often treated equity as a separate set of issues that was unrelated to the development of ambitious instructional practices. Principles of high-quality professional learning call into question the value of PD that is not firmly grounded in high-leverage teaching practices and the instructional materials that teachers are using.

Our findings also raise questions about the value of any instructionally-focused pull-out PD that does *not* explicitly surface issues of equity. It was common to hear in interviews with teachers that while they saw great value in a vision of ambitious teaching, it was not appropriate for their students.[25] For example, in interviews with one hundred teachers in year five of the study, more than half of the teachers expressed sentiments similar to Mr. Dawkins, who suggested that the district's vision and the adopted textbook, *Connected Mathematics Project 2*, was not appropriate for his students because the vision and text "assum[e] that the students can do a level of thinking that they cannot do."[26] It was therefore unlikely that these teachers viewed PD focused on high-leverage teaching practices as relevant. This indicates that it is essential for facilitators to take account of and address teachers' views of their students' mathematical capabilities in all forms of professional learning–and to ensure their design and facilitation supports teachers in developing productive views.[27]

In addition, as described above, teachers reported receiving only minimal support on how to address the needs of students they perceived as struggling in the context of teaching that aimed at rigorous learning goals. It is therefore no surprise that in interviews, teachers often described lowering the cognitive demand of tasks in order to meet what they perceived were the learning needs of their students. For example, in year five of the MIST study, in interviews with seventy-four teachers who described what they did to support students they perceived as struggling, 70 percent of teachers said that they proceduralized instruction, or provided students with a set of steps to take to solve a given problem.[28] As a representative example, Mr. Gomez described telling students "what to do," including providing "notes" on how to solve specific problems, and then having students engage in "independent work" to apply the given procedure to a set of problems. While proceduralizing may result in students getting a correct answer to a problem, in the long term, it does not support students to develop problem-solving capabilities or deep understandings of mathematical ideas.[29] In addition, teachers often reported "taking away" language when working with English learners. This form of practice denies opportunities for English learners to develop language, including mathematical, in the context of mathematics instruction.[30]

We therefore recommend that PD include an explicit focus on how to ensure all students can participate in rigorous mathematical activity, without lowering the cognitive demand of tasks and therefore denying students the opportunity to develop important mathematical understandings, capabilities, and dispositions.[31] Similarly, it is crucial that PD include an explicit focus on how to support English learners in cognitively demanding mathematical activity. This entails supporting English learners to develop language, mathematical and otherwise, *through* engagement in mathematical activity.[32] That said, as we will discuss below in Future Research, the research base regarding how to attend to issues of equity in PD is relatively thin and represents a key area of growth for the field of mathematics education.

Third, pull-out PD must be facilitated by leaders with sufficient expertise in teaching and in supporting teachers' development of ambitious and equitable teaching. Above, we described the challenges that districts face regarding facilitation. Here, we draw on MIST analyses and existing research to identify the forms of knowledge and practice that characterize accomplished facilitators. Our intent is to provide district leaders with guidance in *identifying* people who might be good facilitators. We recognize that these forms of expertise are rare in many districts; below, we discuss how PD facilitators' development might be supported.

As discussed in chapter 4, it is well established in the research literature that pull-out PD facilitators need to be relatively accomplished teachers in the domain in which they are working with teachers.[33] That is, if they are to support teachers to develop ambitious and equitable practices, they need to have developed these forms of practice themselves. As discussed in chapter 3, teaching ambitiously and equitably requires relatively deep mathematical knowledge for teaching, a sophisticated vision of high-quality mathematics instruction and productive views of students' mathematical capabilities.[34] Being an accomplished teacher appears necessary not only in providing technical support to teachers, it also appears important for establishing legitimacy with teachers.

Although being an accomplished teacher is necessary, it is not sufficient to design and lead high-quality PD. In addition, PD leaders need to exhibit what Hilda Borko and colleagues term learning community knowledge.[35] Learning community knowledge refers to the ability to "create a culture of respect, establish group norms, and foster active participation in which teachers share ideas and take intellectual risks."[36] Establishing a community in which teachers are willing to take intellectual risks is particularly important when the goal is to support significant reorganization of their current practices.

In addition, PD leaders need to be able to establish clear goals for teacher learning.[37] That is, they need to be able to assess teachers' current knowledge and practice, and, on the basis of those assessments, determine appropriate goals for teachers' learning, both short and long term.[38] Further, they need to be able to choose and, if necessary, adapt activities to help meet those specific goals for teachers' learning.[39]

Moreover, when enacting the focal activities, they need to be skilled in pressing on teachers' contributions to achieve a professional learning agenda while maintaining a culture of trust.[40] Skilled facilitators attend to teachers' pedagogical reasoning and, in particular, to the rationales that teachers articulate either explicitly or implicitly for their pedagogical decisions. Surfacing these rationales is essential in professional learning settings as it makes explicit not only what teachers would do in a particular instructional situation but also *why* they would do it a focus of discussion and object of inquiry. For example, Sabrina, the expert pull-out PD leader discussed above, regularly pressed on teachers' pedagogical reasoning; this was critical in supporting the teachers' development of ambitious and equitable practices.[41] This kind of press surfaced teachers' uncertainties about mathematical ideas and thus the learning goals for students. In addition, opportunities for teacher learning were greater when Sabrina pressed teachers to make connections between knowledge of students, knowledge of pedagogy, and knowledge of mathematics. For example, in making sense of a complex middle-grades mathematics task, Sabrina pressed teachers to identify the mathematical demands of the task in relation to their knowledge of their own students, and to anticipate where they might struggle. This contrasted sharply with most of the pull-out PD we analyzed in which teachers solved complex mathematics tasks, but were not prompted to identify the mathematical demands as they related to their own students. Other research has shown how through pressing on pedagogical reasoning, a skilled facilitator can surface teachers' unproductive views about their students' mathematical capabilities that influence whether they view ambitious teaching practices as appropriate for their students.[42] Taken together, this research suggests that skilled PD leaders build on teachers' current reasoning to advance a professional learning agenda, much as a skilled teacher builds on students' reasoning to advance an instructional agenda.

Fourth, districts will likely need to support the development of local pull-out PD facilitators. The various aspects of what a facilitator of pull-out PD needs to know and be able to do is daunting. As indicated above, it was rare to find local leaders with these forms of expertise in our partner districts; we anticipate that this is the case in many districts. Thus, districts might choose to seek expertise in facilitation of pull-out PD from outside the district. In all of our districts, central office leaders hired at least some external consultants to design and lead pull-out PD.

However, relying exclusively on external consultants does not build the capacity of district teachers and leaders to engage in informal or formal instructional leadership. Moreover, Paola Sztajn and colleagues' review of the PD literature indicates that external PD programs are only effective in supporting teachers' development of ambitious and equitable practices if they are "adapted to take into account key social and policy features of local settings while maintaining integrity to their core principles."[43] Local PD leaders' knowledge of the district and school contexts is crucial in supporting these adaptations and in establishing a productive learning community for teachers in pull-out PD.[44] Sabrina, for example, as a former math teacher in the district in which she led PD, had a deep knowledge of the local curriculum, pacing guides, and district assessments that the teachers were expected to use. When teachers shared their frustrations with specific district expectations, she was able to empathize with their frustrations and share practical "work arounds" that she had developed when she was teaching.

Therefore, our fourth takeaway focuses on how districts can support the development of local pull-out PD facilitators so that they can support teachers' development of ambitious and equitable teaching practices. This was an explicit object of inquiry in the MIST project.[45] Here, we summarize what we found to be key aspects of supporting the development of local, novice, pull-out PD facilitators. This is consistent with productive designs reported by Hilda Borko and colleagues and Rebekah Elliot and colleagues.[46]

In our work with a set of novice, district-based coaches who were responsible for pull-out PD, we (the research team and central office leaders) engaged in four cycles of work in which we provided PD for the coaches focused on co-planning and facilitating an upcoming PD pilot session with teachers, which the coaches then co-led. The pilot teacher PD sessions were video-recorded so that we could plan for upcoming sessions with the coaches based on an analysis of their current practices, and so that we could engage the coaches in an analysis of their own practice as facilitators. Our analysis of the four cycles of work with the coaches indicates that the coaches increasingly viewed teachers' improvement of their classroom practices as a developmental progression and began to design connected sequences of activities. However, they struggled to facilitate the activities in ways that would meet their ambitious goals for teachers' learning. Although the ways in which the coaches pressed on teachers' contributions differentially did improve, these improvements were not consistent.[47]

A retrospective analysis of the design study described above revealed several principled pedagogical decisions that proved crucial in supporting PD leaders' learning and how we could improve our design for supporting PD leaders' learning.[48] In general, the findings suggest that it is essential that the design of PD for facilitators meets the criteria of high-quality professional learning discussed above and in chapter 4. First, it was important that the coaches engaged in ongoing work together, thereby building the trust to make their facilitation practice an object of inquiry, and to develop a collective framework for planning, facilitating, and evaluating the quality of pull-out PD. Second, it was important that the PD sessions for the coaches were "close to practice" and organized around high-leverage facilitation practices—planning for upcoming PD and pressing on teachers' reasoning in the context of such sessions—that resonated with the coaches as crucial aspects of supporting teachers' learning. A third strength of the design was that the PD sessions for coaches were co-designed and co-facilitated by members of the research team and central office leaders (e.g., head of secondary mathematics, head of curriculum and instruction), who provided complementary forms of expertise. Fourth, the enacted design included coordinated pedagogies of investigation (e.g., analysis of facilitator's practice) and pedagogies of enactment (e.g., co-leadership of pilot teacher PD sessions); both proved important to supporting the coaches' improvement. However, there should have been more opportunities for the coaches to engage in pedagogies of enactment during the coach PD sessions.[49] In hindsight, the coaches would have benefited greatly from opportunities to rehearse their proposed PD activities during the coach PD sessions. As it was, the coaches often proposed activities in their upcoming teacher PD session that seemed promising. However, upon reviewing the video-recordings of the implementation of these plans, we realized that we had not supported the coaches in thinking through the various decisions they would have to make in the moment and the challenges that might arise as they enacted the activities with teachers.

FUTURE RESEARCH

In conclusion, we identify two key areas for future research regarding pull-out PD that we see as essential to supporting instructional improvement at scale. The first concerns attending to issues of equity in the context of pull-out PD. Based on the findings reported above, it appears that the extent to which it is possible to achieve ambitious and equitable teaching at scale depends on the development of robust professional learning designs that support teachers to develop the requisite mathematical knowledge for teaching, instructional vision and forms of teaching practice, *and* to come to see all of their students as capable of engaging in rigorous mathematical activity. At the same time, it is also necessary to support teachers in developing concrete strategies for supporting students who have not historically participated in the intended form of activity. At present, there is little research that focuses squarely on professional learning designs that attend to, simultaneously, these forms of knowledge, practice, and perspectives.[50]

That said, based on a study of MIST teachers who substantially improved their instructional practice over the course of the project, Charlotte Sharpe proposed a curriculum for a sequence of pull-out PD sessions for teachers who exhibit fairly conventional teaching practice that is worthy of future investigation.[51] In brief, she suggested that pull-out PD might initially focus on developing a more sophisticated vision of high-quality mathematics instruction paired with a focus on developing views of one's students as capable of engaging in such forms of activity. This would entail pedagogies of investigation (e.g., analysis of video-recordings of high-quality instruction, analysis of students' reasoning in the context of such instruction as compared with analysis of students' reasoning in the context of conventional instruction). In addition, it is likely important that the pull-out PD is coordinated with opportunities to view one's own students engaged in rigorous mathematical activity. This might take the form of one-on-one coaching, in which a coach models rigorous instruction in a teacher's classroom, while a teacher observes how students they previously viewed as incapable exhibit different capabilities in the context of different instruction. Together, such opportunities to develop one's vision of high-quality instruction and views of students as capable of engaging in and benefiting from such instruction might support teachers to develop reason and motivation to want to work to improve their teaching. Teacher collaborative time and one-on-one coaching might then engage teachers in pedagogies of investigation and enactment that focus on developing the requisite mathematical knowledge for teaching and forms of teaching practice to engage all of one's students in rigorous mathematical activity.

A second area for future research concerns supporting the development of PD facilitators. Improving instruction at scale rests crucially on the expertise of those who support teacher learning.[52] Above, we summarized what the current research base suggests that facilitators need to know and be able to do. Most districts do not currently have a cadre of facilitators with these specialized forms of knowledge and practice. The need for solid designs that support novice facilitators to develop the intended forms of knowledge and practice is acute. We see great value in building on the studies we have cited to investigate how specific aspects of designs support the development of specific forms of facilitators' practice, and how designs need to be adapted depending on particular aspects of a district's context (e.g., number of facilitators, scope of facilitators' work, instructional materials).

SIX

Teacher Collaborative Time: Helping Teachers Make Sense of Ambitious Teaching in the Context of Their Schools

ILANA HORN, BRITNIE D. KANE, AND BRETTE GARNER

IN THE UNITED STATES and abroad, the implementation of teacher collaborative time is on the rise. There is mounting evidence that, in schools where teachers participate in professional collaboration, student achievement may increase.[1] In addition, when colleagues de-privatize problems of practice and reflect together on student learning, they are more likely to sustain their focus on improvement and show instructional growth. For these reasons, teacher collaborative time has become a fixture in the literature on effective professional development for teachers.[2] Around the country, teacher collaboration goes by many names, including *professional learning communities*, *critical friends groups*, and *teacher community*.[3] We use the term *teacher collaborative time* as a general umbrella for the various ways teachers gather with colleagues.

As a part of their instructional improvement strategy, our four partner districts invested in teacher collaborative time by having regular meetings where teams of mathematics teachers could meet and discuss instructional issues. District leaders wanted to provide teachers with opportunities to deepen their content knowledge, design tasks and lessons, and analyze student data. They viewed teacher collaborative time as an important part of the teacher learning subsystem, hoping it would yield and sustain high quality instruction by allowing teachers opportunities to work on school-level issues and adapt what they were learning in pull-out professional development to their particular students and school communities. Despite this intensive push from district leaders for collaboration, we found that *effective* teacher collaboration that had the potential to support teachers' development of ambitious and equitable instructional practices happened relatively infrequently in our partner districts. This finding highlights the very real challenges to establishing effective teacher groups.

In this chapter, we share what we have learned from studying the successful—and the less successful—uses of teacher collaboration time. We begin by describing how collaborative teams were organized in our partner districts and explaining how we studied these teams. We then clarify what we mean by *effective* teacher collaboration, and how we differentiate between more and less effective teacher collaboration. Based upon this definition, we provide examples from one partner district of effective teacher collaborative time. Finally, we describe the challenges to designing, implementing, and supporting effective teacher collaboration before discussing the takeaways regarding how to address these challenges.

UNDERSTANDING THE CONTEXT AND STRUCTURE OF TEACHER COLLABORATIVE TIME IN OUR PARTNER DISTRICTS

The configurations and goals of the teacher groups we studied varied depending on schools' resources (e.g., time, instructional materials, and coaches or other facilitators). Teachers met as a department (e.g., all the math teachers), as grade-level content teams (e.g., all the sixth-grade math teachers), and, less often, as grade-level teams (e.g., all the sixth-grade teachers). They met during or outside the school day on a regular basis, anywhere from daily, weekly, or monthly. The focus of the meetings likewise ranged widely, with some teams focusing on weekly lesson planning, others concentrating on analyzing student data, and still others concentrating on specific instructional practices like introducing or launching rigorous mathematical tasks. A fair number of teams mixed these topics and activities over time. While some teams included a coach or designated facilitator, others did not, though a leader often emerged in these latter teams. Despite these variations in configuration and emphasis, all the teacher collaborative teams we studied met regularly to work on shared tasks of teaching.

How We Studied Teacher Collaborative Time

In years five through eight of the MIST project, we studied how teacher collaborative time contributed to our partner districts' improvement efforts by purposively focusing on the more promising teacher workgroups in the two districts that we worked with for eight years (District B and D). District B mandated teacher collaboration, and nearly all teachers in our sample reported on our annual survey that they met weekly in collaborative teams. In District D, principals were encouraged, but not mandated, to provide common planning time for mathematics teachers. Nonetheless, nearly all teachers in our study reported participating in teacher collaborative time. They typically met weekly, either during the school day, before students arrived, or after student dismissal.

We used an approach called internal sampling to select teacher groups to study, and asked key district coaches and math specialists to recommend teacher teams that collaborated well.[4] We planned to follow these groups for four years. However, we found it necessary to select focal groups each year because the groups were frequently reconfigured for a number of reasons, including high rates of turnover within teacher groups. Over the course of the study, fourteen groups left or were dropped from our sample. Seven groups were dropped because, despite their strong reputation as a recommended group to follow, we saw little evidence of deep engagement in the collaborative work. The remaining seven left the study because of changes in school staffing, as key instructional leaders left the school or the entire staff was reconstituted. This turnover was prevalent, since teachers often changed schools or grade levels across years. In the end, we observed a total of twenty-four workgroups, video recording between four and six meetings from each group per year, for a total of 111 meetings. We supplemented the video data with other data from the MIST project, including the annual surveys and semi-structured interviews of district leaders, principals, coaches, and teachers, where we asked how teacher collaborative groups were organized, typical activities for teacher groups, the helpfulness of these activities, and how they saw the activities in teacher group meetings influencing their own or others' instructional practice.

THE PROMISE OF TEACHER COLLABORATION: WHAT MAKES IT EFFECTIVE?

It is worth elaborating on what we mean by *effective* teacher collaboration. Teachers often find any kind of collaboration useful, since it breaks professional isolation, allows them to share the burden of the numerous tasks they must complete, and might even provide emotional support by offering adult ears for situations mostly witnessed (and judged) by young people. When we assess the effectiveness of collaboration, we distinguish between collaboration that helps teachers carry out instruction as usual and collaboration that stands to support the development of ambitious and equitable instructional practices, with our interest primarily on the latter.

Prior studies indicate that teacher collaborative time is most effective in supporting teachers' learning about ambitious and equitable instruction when it involves work that is structured, focuses on student learning, and is connected to teachers' daily practice. Ideally, meetings should be forums for sharing expertise and resources that support teachers' sense-making. At its best, teacher collaborative time can provide teachers with opportunities to contend with school-level problems of practice and adapt the big ideas from pull-out professional development to the complex daily realities of particular classrooms.[5] For instance, a professional development provider might suggest that teachers ask more open-ended questions to engage students in a lesson ("What do you notice or wonder about this problem?") rather than the closed-ended questions that are typical in procedural mathematics classrooms ("What's the next step?"). A teacher might then try out this suggestion in her classroom, adjusting her questioning strategies, only to find that her students generate unexpected responses, including novel solution paths and off-topic commentary. While the teacher might agree in principle about incorporating open-ended questions, she needs support to make the practice work in her classroom: adjusting the wording of questions to elicit student thinking, providing appropriate conceptual supports, or deciding which student observations might lead to a fruitful discussion. It is not feasible to address all these detailed decisions in pull-out professional development. Teacher collaborative time—particularly under the guidance of an expert facilitator—can provide important spaces for teachers to discuss these problems of practice and think through different adaptations of ambitious and equitable instruction.[6] However, for collaborative time to support teachers in trying out and adapting particular practices, the team needs to develop a sense of trust and mutual support, as well as ways to reconcile different perspectives.[7]

Understanding how to take a big idea about ambitious math instruction and tailor it meaningfully to their school and classroom setting involves knowing not only the *how* of practice, but understanding the *why*, with the latter allowing teachers to make adaptations that preserve the integrity of a teaching practice. For instance, a teacher may succeed in getting students to talk more by incorporating more open-ended questions in her classroom. In that case, she is successful in *how* she uses the practice. However, she may be less effective at connecting students' contributions with important mathematical ideas, which is one of the key *why's* of the practice. Without strong guidance on tailoring and adapting new instructional practices—a key function of effective teacher collaboration—teachers might not meet the goals of ambitious math teaching.

CATEGORIES OF EFFECTIVE TEACHER COLLABORATIVE MEETINGS

As part of our investigation, we categorized teachers' collaborative meetings in terms of the opportunities for teachers' learning that they provided. In an ideal world, teacher groups would discuss both the *how* and *why* of instruction: When teacher groups did so, we coded these discussions as supporting greater opportunities for teacher learning. Table 6.1 shows the six categories of teacher collaborative meetings that we identified, ordered from least to most effective in terms of the extent to which the conversations addressed the *why* and *how* of instructional decisions. The thick horizontal line separates meetings that do not explicitly address *why* from those that do.

In meetings that are the least productive in terms of teacher learning opportunities, teachers have *conflicting goals,* and there is no focus on either the *why* or the *how* of instructional decisions. For instance, in a *conflicting goals* meeting, one teacher may hope to schedule an upcoming common assessment during the meeting, while another may expect to disseminate work associated with lesson planning for following week. In fact, we did

TABLE 6.1 Categories of teacher collaborative meetings, ordered from least to most effective in terms of teacher learning opportunities

CATEGORY	FOCUS ON WHY	FOCUS ON HOW
Conflicting Goals	No	No
Pacing	No	Yes: Pace of instruction coordinated
Logistics	No	Yes: Pace and topics of future instruction coordinated
Tips and Tricks	No	Yes: Instructional talk or activities for future instruction coordinated
Collective Interpretation, Separate from Future Work	Yes: Teachers discussed purpose of instructional decisions	No
Collective Interpretation, Linked to Future Work	Yes: Teachers discussed purpose of instructional decisions	Yes: Instructional decisions linked to future instruction

not identify any conflicting goals meetings in our sample, likely due to our selection of workgroups. However, such meetings have been documented in other studies.[8]

Pacing and logistics meetings accounted for more than 40 percent of the meetings in our sample (see figure 6.1). The distinction between these two meeting types was subtle. In *pacing* meetings, teachers set their calendar for when specific lessons might be taught, but mathematical topics were not discussed (e.g., "We'll teach lesson 4.2 on Monday, but we need two days for 4.3"). In *logistics* meetings, there was sometimes more explicit consideration of mathematical content, since the topics themselves were named (e.g., "We'll review one step equations on Monday, but we will need two days to introduce two-step equations with integers"). Both of these types of conversations help teachers identify *how* they plan to teach different lessons but do not discuss the *why* of these instructional decisions. That is, teachers discuss when they plan to assign particular tasks and activities, but they do not discuss a rationale for these activities in terms of how they stand to support student learning.

In *tips and tricks* meetings, which were about one-quarter of the sampled meetings, teachers share small glimpses of their practice for the lesson they

FIGURE 6.1 Frequency of the different types of teacher collaborative meetings. CI-FW refers to the category *collective interpretation separate from future work* and CI+FW refers to the category *collective interpretation linked to future work.*

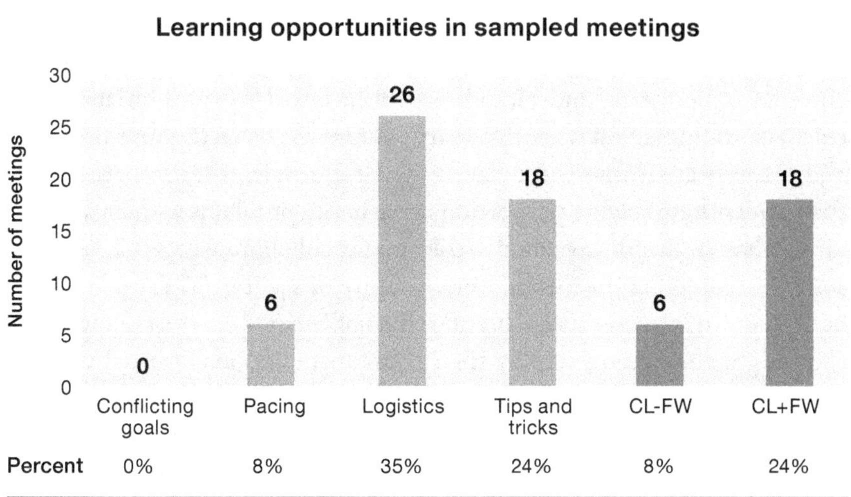

are planning. For example, teachers might share mnemonic devices for teaching a particular topic or provide a quick instructional explanation ("When I teach this lesson, I always tell them . . ."). The learning opportunities are richer in tips and tricks meetings than in pacing or logistics meetings, since teachers get examples of instructional practice to take back into their own classroom. However, these meetings are not as rich as the later categories in our taxonomy, because tips and tricks meetings rarely involved a back and forth discussion between participants, which limited teachers' opportunities to discuss the *why* of practice. Again, these meetings focused primarily on the *how* of teaching, with little discussion of why these strategies helped meet instructional goals.

Because we identified teacher teams that collaborated well, about one-third of our meetings involved some discussion of *why*, what we call *collective interpretation.* In these conversations, teachers might identify a problem they are working on in their instruction and weigh different ways of addressing it. As they consider various paths, they exchange ideas about the trade-offs of each response. In doing so, they not only have a chance to consider their own thinking but also to hear from their colleagues. This focus on the *why* of teaching supported greater opportunities to learn about ambitious and equitable instruction.

There were two sub-types of collective interpretation: *not linked to future work* and *linked to future work*. In both types of conversations, teachers might engage in collective interpretation as they analyze student work samples and delve into the *why* of students' responses, investigating their thinking and current mathematical understandings. In the *linked to future work* meetings, teachers would harvest what they learned from the collective interpretation and incorporate it into their planning. For example, if teachers recognize that students are solving proportional reasoning problems with an additive rather than multiplicative model, they might collaborate together on a lesson that highlights the multiplicative nature of scaling. (This example will be elaborated later in this chapter.) In the *not linked to future work* meetings, they might then go on to plan their subsequent lessons as usual without considering the insights yielded by their initial sense-making discussions. Although both kinds of meetings support teachers' investigations into the *why* of practice, only the first type of meetings support an explicit connection to future instruction, thereby providing greater opportunities for learning.

ILLUSTRATIONS OF EFFECTIVE TEACHER COLLABORATION

To illustrate effective teacher collaboration, which involves collective interpretation in the final two categories in our taxonomy (see table 6.1), we will anchor our discussion in two examples from District D, the Aspen Middle School and Magnolia Middle School sixth-grade math teams, two schools who had multiple meetings where teachers engaged in *collective interpretation linked to future work*. Magnolia Middle School's sixth-grade team had one of the strongest collaborations in our sample. They focused their collaborative time on understanding and responding to students' mathematical thinking. They met weekly and were led by experienced instructional leaders, Coach Lindsay and Assistant Principal Mr. Donovan. During their meetings, the Magnolia team worked across a three-week cycle.

In the first week, they used student assessment data to identify troublesome math topics. They selected a standard with which many students had struggled on in an interim assessment, analyzed it together to come to a shared understanding of what "proficiency" looks like on the standard, and developed a common formative assessment to give in the coming week.

In the second week, the group analyzed student work on the common assessment. Led by Coach Lindsay and Mr. Donovan, both of whom had substantial math teaching experience and a strong understanding of ambitious instruction, the teachers made sense of their students' thinking and designed instructional strategies to address common errors. In the third week, the group reported back on their implementation of the previous week's instructional strategies and assessed their effectiveness by analyzing new work from a handful of students who were just on the cusp of mastery. They then repeated this cycle of information gathering, analysis, and instructional response to make sure they addressed students' sense making in their instruction.

Throughout the cycle, Coach Lindsay and Mr. Donovan facilitated rich discussions. To do so, they solicited detailed descriptions of the teachers' classrooms and practice as they looked at the student work. Their discussions about students' mathematical understandings oriented teachers toward students as sense-makers. When the teachers co-designed common assessments, the facilitators pressed them to articulate rationales for instructional decisions that are tied to ambitious and equitable goals for student learning.

Aspen Middle School's sixth-grade collaborative math team also stood out for its strong collaboration. A highly knowledgeable math coach, Heidi, facilitated the group. In two of their meetings, the group focused on improving practice by working to design and rehearse introducing or "launching" rigorous mathematical tasks so that all students could begin working on them productively, a key practice of ambitious and equitable instruction. During a launch, teachers ideally maintain the cognitive demand of tasks as they discuss and develop common language for describing key contextual features and key mathematical ideas in a problem.[9] Coach Heidi's approach was quite promising: The teachers took turns role-playing launches, which set up opportunities for her to give feedback focused primarily on the *how* of this practice. In addition, Coach Heidi provided opportunities to discuss both the *how* and the *why* of launches, reminding teachers that a good launch should result in students being able to start a task using a variety of strategies. In interviews, the Aspen teachers reported great trust in Coach Heidi, and this was evident in their engaged participation in the activities she led.

These two examples highlight different strengths and possibilities for teacher collaborative time. Both Coach Lindsay and Coach Heidi were highly knowledgeable in ambitious and equitable instruction, and they used routines (*the three-week assessment cycle* and *role-playing rehearsals*) to guide teachers' learning during collaborative time, offering site-specific forums for professional learning.[10] Also, at both schools, teaching was de-privatized as teachers shared and discussed student work samples or rehearsed launches. Additionally, the Aspen team's trust in Coach Heidi's leadership was a critical asset for sharing practice.

Notably, both teams centered teachers' attention on issues of student learning, something research has repeatedly shown to matter in effective teacher collaboration. At Magnolia, the three-week assessment cycle routine squarely focused on evidence of students' understandings and responding to student thinking through instruction. For example, the group at Magnolia made sense of mathematical topics that students found challenging by analyzing students' thinking on an informal assessment the group designed around unit rate. Then, based upon the group's findings about students' mathematical thinking, the two facilitators at Magnolia introduced a new instructional strategy—in this case, a double number line—that teachers might use to support students in making sense of the difference between additive and multiplicative problems.

At Aspen, when teachers co-designed launches, Coach Heidi asked teachers to look at math tasks and predict where students might encounter challenges. The Magnolia routine focused on examples of student struggle, while the Aspen routine focused on anticipated student struggle. Both kinds of analysis of student thinking are crucial to supporting teachers' development of ambitious and equitable practices that build on students' ideas. As described in chapter 7, there is evidence that analyzing student work, considering student thinking, and investigating and enacting new instructional strategies, such as the double number line or the launch, have potential to support teachers' professional learning.

Reporting on specific examples of the mathematical challenges that students encountered allowed the Magnolia teachers to share experiences and solicit input on common problems of practice, giving them the opportunity to develop new ways to think about critical instructional problems.[11] Coach Lindsay and Mr. Donovan pressed hard to keep the group's focus on core issues of teaching and learning, and connecting student learning to instructional issues is a key characteristic of effective teacher collaborative time.[12] Additionally, these discussions made the challenges in student learning a normal part of teaching—they are not anyone's fault, but rather a part of the learning process. This helped teachers move away from deficit language when students encountered difficulty. That is, instead of merely saying that students who encounter mathematical challenges are "slow learners," the teachers analyzed the students' current thinking and then discussed how to respond instructionally. For example, in looking at a specific work sample, teachers might determine that a student used unit rate to determine the amount of money earned after a certain amount of time, but he did not demonstrate sufficient conceptual understanding because he relied on an additive relationship to complete a table in a word problem.[13] This detailed understanding of their students' thinking pointed teachers toward a course of action: to address students' conceptual understanding of unit rate. In the process, teachers moved toward more productive framings of students struggling to solve challenging tasks.[14]

These conversations were not always straightforward. At times, the Magnolia teachers were perplexed by students' errors, and they were often surprised that their instruction had not adequately addressed key ideas. However, seeing evidence that other teachers' students were also encountering mathematics challenges helped them to be less defensive about the quality of their instruction and move toward actionable solutions for their next lessons. When teachers reflect on their students' learning and their practice together, it can demonstrate that teaching challenges are normal, while also providing emotional support to persevere despite the uncertainty that comes with innovation.[15]

CHALLENGES TO EFFECTIVE TEACHER COLLABORATION

As we have already noted, unlike Aspen and Magnolia Middle Schools, not all teacher collaboration met our standards for effectiveness. Even though we sampled for strong collaborative teams, the low yield of effective meetings is not surprising. Prior research documents a number of challenges that impede effective collaboration. These include: a lack of resources (i.e., time to meet, money to compensate teachers' time per district contract, access to rigorous tasks, and high-quality facilitation); existing workplace norms such as privacy (i.e., teachers not wanting to share problems of practice), autonomy (i.e., a professional ethos of "to each their own"); taken-as-shared deficit language for talking about students; as well as other initiatives and collaborative activities that compete for teachers' attention.[16] We saw many of these and additional issues at play in our partner districts.

In addition to these well-documented obstacles, our work highlighted a number of other challenges to effective teacher collaborative time. As was described in chapter 4, coordinating team goals with the larger teacher learning subsystem is the first challenge in achieving effective collaboration. When pull-out PD, coaching and teacher collaboration are aligned, improvement initiatives have a greater chance of taking root in teachers' practice. Unfortunately, this happened infrequently in our sampled meetings. District-level initiatives, such as District D's focus on launching cognitively demanding math tasks, seldom found their way into teacher collaborative meetings. (Aspen was the only group in our sample of teacher collaborative teams that worked on this practice.) The few times we saw coaches bring in ideas from pull-out professional development, teachers often went along with activities begrudgingly, reporting that it did not seem relevant to their urgent daily work.[17] This finding reminds us of the findings of prior studies that emphasize that goals for collaborative time should be set in collaboration with teachers.[18]

A second challenge of particular prominence was allocating teachers' time for collaborative work. In our most effective teacher work groups, collaborative time was scheduled during the teachers' contractual workday *in addition to* teachers' regularly scheduled planning periods. This is important, because effective collaborative work is time consuming. For instance, at Magnolia Middle School—our most effective team—teachers said that the collaborative time was useful to their practice, but they also highlighted that it was difficult to complete what they often saw as "extra" work. At Magnolia, the collaborative work involved preparatory activities, such as collecting and analyzing student work in addition to participating in weekly collaborative meetings. These tasks had to be added to a list already crowded by more traditional tasks of teaching: lesson planning, making copies, grading student work, communicating with parents, attending IEP meetings, and the like.

Third, high rates of teacher turnover and policy churn, which we witnessed, affected workgroup continuity from year to year, jeopardizing the crucial trust at the heart of meaningful collaboration. The year after we followed the Magnolia team, Coach Lindsay moved and other teachers left the school. Likewise, Coach Heidi's work as an instructional coach was waylaid when school leaders shifted her role to school data manager.

Fourth, there was an absence of clear leadership and instructional expertise in many of the meetings we observed, resulting in inconsistent press on issues related to ambitious and equitable teaching. Important topics went unexamined, leaving instruction fundamentally unchanged. For example, in one meeting, the teachers realized that students were confusing the volume formulas for spheres and cylinders, but concluded that explaining the steps again would address any misunderstanding, thus leaving out deeper connections between these formulas and the structures they describe. Generally, in less effective teacher collaboration, instead of using colleagues to troubleshoot and adapt practices to their particular students, classrooms, and communities, teachers focused on other pressing tasks, such as curricular logistics, allocating resources, and coordinating calendars, which is unlikely to serve the transformational goal of district-wide instructional improvement.

SCHOOL LEADERS' CONTRIBUTIONS TO COLLABORATION

School leaders shaped the effectiveness of the teacher collaborations in our sample. They supported effective collaboration in several ways, depending on their own expertise and experience. First, by furnishing teacher teams with facilitators who had instructional expertise, school leaders helped ensure that conversations would center on vetted ideas about ambitious and equitable mathematics instruction. In some cases, this required leaders to seek an expert coach outside of the school. In other cases, leaders appointed a skilled teacher to take on the facilitation role in the meetings. If school leaders themselves had knowledge of ambitious mathematics teaching, as with Magnolia, they could design and lead effective teacher collaboration. When this expertise was not available or removed from the team, this greatly diminished teachers' learning opportunities. At Aspen, described above, school leaders eventually asked Coach Heidi to become the school's Data Manager, effectively eliminating her position as an instructional leader and greatly weakening the teacher collaboration.

Second, school leaders supported effective collaboration by protecting teachers' time during appointed meetings. In our strongest groups, school leaders emphasized the importance of teachers using the time for meaningful planning and inquiry, thereby supporting their ongoing learning. Given the countless administrative tasks school leaders are asked to see through to completion, they are often tempted to assign some of these tasks to teachers during their "free" time. In this way, time-consuming logistical tasks–such as textbook distribution, district reviews, or linking lessons to standards–can infiltrate collaborative meetings, moving the focus away from student learning. Some of the focal groups we dropped from our study spent considerable periods of time on such tasks, giving them little time to discuss the substance of their teaching.

Finally, school leaders who had a *professionalizing stance* on their teachers' collaborative work helped to inculcate the trust and mutual support necessary for teachers to share their problems of practice.[19] That is, they viewed their teachers as competent professionals who, with reasonable support, could be trusted to focus on improving their mathematics instruction in the support of students. In contrast, at one of our schools, school leaders took a *surveillance stance* on their teachers' collaborative work, sitting in on workgroup time to steer conversations away from instructional issues towards state accountability testing. Not only did this lower the quality of the teachers' workgroup conversations, but it also fragmented the department, leaving teachers less inclined to seek informal help from colleagues in ways that could improve their instruction.[20]

KEY TAKEAWAYS FOR THE DESIGN AND FACILITATION OF TEACHER COLLABORATIVE TIME

As with other facets of the teacher learning subsystem, the value of teacher collaborative time depends on its focus, the quality of the activities in which teachers engage, and high press facilitation. Drawing on the "Design Principles for a Potentially Productive Teacher Learning Subsystem" (chapter 4), we offer the following to characterize effective teacher collaboration time (see table 4.1)–with the caveat that teacher collaboration, on its own, is unlikely to support teacher learning.

Ensure that the focus of teacher collaborative time is connected to pull-out professional development and other instructional supports. The focus of various supports for teachers' learning (e.g., district-wide pull-out professional development, teacher collaborative time, content-focused coaching) should be coherent and tightly connected so that the goals for improving classroom practice being worked on in one type of support are built on and elaborated in other types of support. At its best, pull-out professional development helps teachers engage with big instructional ideas, while high-quality instructional coaching helps teachers work on specific practices with particular groups of students. In a well-articulated teacher learning subsystem, teacher collaborative time offers something in between: Teachers can make sense of big instructional ideas, but they can also hear how other teachers in their school are adapting them to their classrooms, giving them a better repertoire for meeting the needs of different students. In this way, they are not learning general practices or specific tricks, but are instead learning how an instructional approach can be adjusted to support the learning of diverse groups of students. This allows teachers, as learners, to develop deeper understandings of both the *how* and *why* of ambitious and equitable instruction.

Coach Heidi connected pull-out professional development on launching cognitively demanding math tasks to teacher collaborative time. In designing launches for upcoming lessons and rehearsing them, the teachers had opportunities to understand *how* to enact the launches. The discussions they had about the core mathematical ideas that were the focus of the launch rooted their activity in the *why* of the launch: to enable students to begin working productively on tasks. In one meeting that focused on designing a launch, the teachers analyzed a mathematical task to anticipate ways students might be confused. Some teachers proposed that place value would be a stumbling block, while another thought that exponential notation might prove problematic. After substantial debate, they agreed that the crucial understanding for the problem lay in students' understanding of zero and, therefore, that this should be the focus of their launch. When teachers enacted the launch in their classrooms after the meeting, they could bring that understanding to bear, focusing their questioning and feedback to students on that core mathematical idea. Despite this encouraging example, instances in which teacher collaborative time connected to other facets of the teacher learning subsystem were infrequent. Even strong teacher teams did not necessarily align their work to pull-out professional development, perhaps because they did not see how the professional development connected to what they viewed as urgent instructional problems.

Instructional coaches or teacher leaders with expertise in both ambitious teaching and supporting teachers' learning should facilitate teams. Facilitators of designed supports for teacher learning should have expertise in ambitious and equitable teaching, as well as practical knowledge of teacher development (see chapters 5 and 7). Facilitators like Coach Lindsay, Mr. Donovan, and Coach Heidi not only engaged teachers in potentially productive activities, but they continually pressed teachers to refine their understanding of key ideas in ambitious and equitable mathematics teaching. As we saw in the Magnolia three-week cycle example, strong facilitators (1) solicited detailed descriptions of teachers' classrooms and practice, (2) oriented teachers toward students as sense-makers, and (3) pressed for teachers to articulate rationales for instructional decisions that are tied to ambitious and equitable goals for student learning.[21]

Skilled facilitators not only design productive learning activities for teacher collaborative time, as happened at Magnolia and Aspen, but they also hold teachers accountable to explore and explain their reasoning—what we have come to call *facilitator press.*[22] Specifically, good facilitators often follow up teachers' contributions with questions that encourage them to elaborate their thinking. This not only clarifies the pedagogical reasoning of the teacher; facilitator press also helps the other teacher understand the reasoning. For example, in one exchange between Coach Lindsay and Deanna, a teacher, Deanna asserted that a student's work showed that he "has the concept." Coach Lindsay pressed her to clarify: Does he have the concept or the procedure? Deanna looked at the work again for evidence, realizing that although the student had developed a procedure, the student's understanding of the underlying idea might have been fragile. Not only did this exchange support Deanna's learning, but the other teachers had opportunities to clarify the important distinction between being able to recreate a procedure and understanding a concept. Pedagogical reasoning, when made visible through the press of a skilled facilitator, gets the critical *why* of instruction into teachers' conversations. This type of dialogue characterized the richest teacher collaborative meetings.

Develop a shared long-term goal for instructional improvement and align activities with that goal. As we stated above, activities during teacher collaborative time should align with broader improvement goals that support teachers' development of ambitious and equitable mathematics instruction. The facilitator should support teachers in developing shared goals for their collective work, and in developing activities and routines that help them make progress toward their goals.

Goal setting turns out to be a central component of any initiative to improve student learning.[23] Crucially, goals should be negotiated with teacher teams to develop a sense of shared purpose. Strong facilitators of teacher collaborative teams must contend with the tension between meeting teachers' needs and pressing on aspirations for improvement.[24] Finding the sweet spot between team leaders' and teachers' aspirations may require some time, but many possibilities exist, such as facilitating deeper mathematical discussions or incorporating productive assessment within each unit of instruction. Likewise, when developing goals, facilitators must also be mindful of schools' resources and culture.[25] By identifying goals that are meaningful to the teachers themselves, facilitators can foster buy-in and trust among teacher teams.

Although the day-to-day task of planning the *pacing* and *logistics* of lessons are a common focus in teacher collaboration, effective teacher groups continually press on larger goals for improvement by linking these activities to the *why* of their instructional decisions.[26] A larger purpose moves teachers past the pressing urgency of daily tasks and toward a longer view of instruction; instead of worrying only about getting through a day or week, teacher workgroups can set goals for several weeks, a semester, or a year. This different perspective can help them gauge their work and its relationship to their own classrooms beyond daily goals of lesson planning toward longer term goals rooted in richer visions of mathematics teaching and learning.

Schedule time during the contractual workday for teachers to work with the same colleagues in a sustained way. In order to develop a truly collaborative workgroup, teachers should have adequate time during the school day to work on problems of practice with the same group of colleagues across multiple years. Although the strongest collaborations we observed were disrupted by staffing and policy churn, there are examples in the research literature that illustrate the benefits of sustained collaboration. When the same group of teachers work together over time, they have opportunities to develop great trust, strong norms for sharing practice, a common vision for high-quality mathematics teaching, routines for investigating problems of practice and for inducting new team members, productive ways to disagree, and institutional knowledge about advocating for their professional needs.[27]

Sustaining effective teacher collaborative time was a persistent challenge for our partner districts. Teacher turnover was an on ongoing challenge across the districts, which made it difficult for workgroups to develop strong collaborations. Teachers were frequently moved across grade levels or changed schools (or left the districts altogether), and coaches' responsibilities were frequently changed, which limited opportunities for long-term collaboration and professional development. Our hunch is that the teams can sustain some staffing changes, but trust and flow of their work are disrupted when a team loses a particularly knowledgeable member or finds itself with fewer than half of the original participants.

Finding time for collaboration was another persistent challenge. Even when teachers' and coaches' assignments remained relatively stable, some groups opted to meet very briefly (if at all) each week. In the context of teachers' work, this unfortunately made sense, especially when the common planning time took the place of individual planning time and was thus viewed as additional work on top of existing responsibilities. We suggest that time be set aside for collaborative work *in addition to* teachers' regularly scheduled planning periods. It may also be prudent to attend to teachers' workload in other ways; for instance, school leaders might reduce the number of different "preps" for which teachers are responsible whenever possible.

Leverage opportunities to meaningfully adapt instructional strategies to particular students' and schools' resources and needs through pedagogies of investigation linked to pedagogies of enactment. The activities in which teachers engage should be close to instructional practice and organized around high-leverage aspects of teaching practice. Using pedagogies of investigation and pedagogies of enactment, teacher collaborative time should support teachers in understanding the *why* and *how* of ambitious mathematics instruction. Recall from chapter 4 that teacher learning requires a balance of pedagogies of enactment and pedagogies of investigation.[28] *Pedagogies of enactment* describe how teachers learn by trying out some of the core practices of teaching with opportunities for feedback from expert colleagues. *Pedagogies of investigation* refer to the ways teachers can inquire into their practice. For example, the Aspen team's activities used both approaches. The *enactment* aspect of the team's launch rehearsals is apparent: they practiced designing and doing launches. At the same time, the discussions Coach Heidi led around the design and enactment decisions supported a collective *investigation* of this practice. Notably, the practices that Coach Heidi and her team investigated and enacted were at the heart of ambitious and equitable mathematics teaching, which seeks to support the development of learners' thinking as they tackle rich mathematical tasks.

The need to balance pedagogies of enactment and pedagogies of investigation echoes our earlier observation that teachers' learning opportunities are best fostered when discussions focus on both the *how* and *why* of practice. By trying out valued forms of practice in a team setting, teachers can provide both models for and feedback to one another. We found that pedagogies of enactment and investigation were tightly coupled in our strongest groups.

Across our sample of workgroup meetings, we found that most teachers had few opportunities to engage in activities that connected pedagogies of enactment and investigation. Importantly, not all enactments count as pedagogies of enactment. In the form of collaboration that we refer to as *tips and tricks* (see table 6.1), we saw several teacher teams model lesson launches without giving one another feedback. In these groups, teachers took turns modeling launches (often by saying "This is how I'm going to start . . .") but did not discuss the differences in their approaches or the reasoning behind their choices. Although teachers surely got ideas from their colleagues' launch examples or specific wordings for instructional explanations—the *how* of practice—the lack of dialogue around their choices limited teachers' learning by leaving out the *why*.

DIRECTIONS FOR FUTURE RESEARCH

From this chapter, it should be clear that supporting the development of productive teacher collaboration is as infrequent as it is potentially useful. While instructional improvement has been repeatedly shown to co-occur with strong teacher communities, it remains unclear how to foster and sustain a sense of collective responsibility for students' learning. Researchers and practitioners interested in instructional improvement at scale would benefit from future research in three related areas. First, we need a clearer understanding of the processes and conditions that help teacher collaborative time contribute to equitable outcomes for students. That is, how does effective teacher collaboration come about? In what conditions is it sustained? Reading earlier research and looking at our own data, we suspect that the frequently observed positive relationship between teacher collaboration and greater-than-expected student achievement is an important consequence of productive and well-facilitated teams.[29] In other words, such teams sustain teachers' engagement in instructional improvement projects and support them in continually looking at the possibilities within their students and community for engagement and learning. The Magnolia team's three-week assessment cycle helped teachers connect instruction to students' current mathematical understandings.

A second related area of research centers on how effective teacher collaboration can support the development of productive views of students' current mathematical capabilities, which rely on teachers having productive instructional responses to students' learning difficulties. Studies that investigate whether and how different types of collaborative activities support improvements in teachers' views of their students' capabilities could help explain why some teacher teams positively influence student learning, while others do not manage to do so.

Finally, tying these two lines of research together, we would similarly benefit from a deeper understanding of effective facilitation. What kinds of knowledge and understandings do skilled facilitators bring to their work? We see that our best facilitators have strong mathematical knowledge for teaching and sophisticated visions of high-quality mathematics instruction. In addition, like our best instructional coaches, they appear to understand something about teacher development. Pragmatically, how do facilitators effectively manage tensions between supporting teachers with the immediate and pressing tasks of teaching with broader aspirations to improve instruction? Teachers seldom have enough time to do all the tasks of teaching, ambitious or not. Asking them to use their time to engage in something like looking at student work or rehearsing launches may seem "non-essential," yet we see that some facilitators managed to persuade their teams that these activities were worthwhile.

SEVEN

Understanding Content-Specific Instructional Coaching: On-the-Ground Support for Teacher Development

BRITNIE D. KANE, PAUL COBB, AND LYNSEY GIBBONS

INSTRUCTIONAL COACHING is one of the fastest growing forms of instructional support for teachers.[1] A number of studies provide evidence that instructional coaching initiatives can support improvement in teachers' instructional practice, as well as students' performance on achievement tests.[2] The substantial promise of instructional coaching is that coaches can act as more accomplished colleagues who ground teachers' development of ambitious and equitable instructional practices in the specifics of their daily work with their students in their classrooms.[3] In other words, coaching has the potential to support teachers in learning "from practice in practice."[4] In line with other research on coaching, we define an instructional coach as a person with content-specific expertise hired, full or part time, to support teachers to improve the quality of their instruction.[5] Thus, we include teacher leaders in our definition of *instructional coach*.[6] We will use the term coach in this chapter with the understanding that it encompasses both coaches and teacher leaders.

Despite the substantial promise of instructional coaching initiatives, findings about the extent to which instructional coaching supports improvements in teachers' instructional practices have been mixed.[7] The literature on coaching highlights a number of challenges for the implementation of coaching initiatives. For instance, instructional coaching requires substantial expertise, yet there is evidence that many coaches may still be developing the intended instructional practices themselves.[8] Also, the teaching profession in the United States has traditionally emphasized professional egalitarianism and privacy, which can make it difficult for coaches to earn teachers' trust and gain access to their classrooms.[9] Finally, coaches' roles are often varied and vaguely defined, and they are frequently called upon to do work unrelated to supporting teachers to improve their classroom instruction.[10]

Consistent with other work on instructional coaching, our partner districts implemented a range of different content-specific coaching programs, and their coaching initiatives shifted over the course of our eight-year collaborations with them.[11] For example, in one of our partner districts, coaches were hired during the same school year either by district mathematics specialists in charge of coaching or by principals at individual schools. The coaches hired by district mathematics specialists were expected to work with mathematics teachers in several schools and were accountable to the mathematics specialists, whereas the coaches hired by principals worked full-time in a single school and were accountable to their principals.[12] In another district, middle-school principals each hired a mathematics teacher from the current teaching staff to act as a mathematics coach for part of each day.[13]

Against this backdrop, this chapter focuses on how instructional coaches can contribute to mathematics teachers' development of ambitious and equitable teaching. MIST investigations of coaching aimed to identify the types of activities in which coaches should engage teachers to support their learning, to clarify why some coaches but not others were able to enact these potentially productive activities effectively with teachers, and to understand aspects of school and district contexts—with particular attention to the role of the principal—that influence the extent to which coaches are able to work with teachers to support their development of ambitious and equitable practices. In the concluding section, we offer key takeaways for practice and make suggestions for future research. Although instructional coaching is a core facet of a teacher learning subsystem, our findings highlight that coaching initiatives require careful design and implementation if they are to support teachers' development of ambitious and equitable instructional practices.

IDENTIFYING POTENTIALLY PRODUCTIVE COACHING ACTIVITIES

As we have noted, instructional coaching promises to provide teachers with ongoing support from a colleague who has already developed relatively accomplished instructional practices. However, making good on this promise relies on knowing which types of coaching activities have the potential to support teachers to improve the quality of their instruction. In addition, identifying potentially productive coaching activities can clarify how coaches' roles should be defined and guide decisions about how coaches might focus their efforts to support instructional improvement. Unfortunately, the research base on this issue is thin and provides only limited guidance. As a first step in addressing this issue, we therefore drew on research on teacher learning and on high-quality professional development to identify criteria for potentially productive coaching activities.[14] This synthesis of prior research indicated that potentially productive activities for supporting teachers' learning:

- are sustained over time
- are close to daily practice
- focus on high-leverage aspects of instruction
- foreground students' thinking
- engage teachers in both investigating and enacting ambitious and equitable forms of practice[15]

Notably, these ideas are closely related to the principles for a potentially productive teacher learning subsystem that were laid out in chapter 4.

As a second step, we searched the coaching literature for activities that both (1) met all five of the above criteria and (2) there is empirical evidence that they can support teachers' development of ambitious and equitable instructional practices. We call the resulting coaching activities *potentially* productive rather than productive to acknowledge that these activities will not support teachers' learning unless they are enacted effectively. In sharing the findings, we distinguish between activities that coaches might use when working one-on-one with teachers in their classrooms and those that they might enact with groups of teachers.

WORKING ONE-ON-ONE WITH TEACHERS IN THEIR CLASSROOMS

We found evidence in the current literature that the following three one-on-one coaching activities can support teachers' development of ambitious and equitable practice: (1) modeling instructional practice, (2) co-teaching, and (3) the coaching cycle.[16] All three activities capitalize on what has been called instructional coaching's "most distinctive feature:" that it is grounded in the specifics of teachers' daily work with students.[17] This grounding enables coaches to discuss core principles of ambitious and equitable instruction with teachers "not in a vacuum but in the context of practice."[18] We also briefly discuss observing and providing feedback because it is a very common one-on-one coaching activity even though the research evidence indicates that its potential to support teachers' learning is questionable.

Modeling

In modeling, a teacher observes a more accomplished colleague, often a coach, enact particular instructional practices in the teacher's classroom. A number of studies of instructional coaching and teacher education highlight that modeling can support teachers in developing an image of the accomplished enactment of instructional practices.[19] As one teacher educator put it, modeling provides a "living example" of instructional practice.[20] Although there is more to learn about how coaches might enact modeling most effectively, we speculate that modeling may be particularly effective as an early step in helping teachers to understand new instructional practices.[21]

There are also indications that modeling may be especially valuable for supporting teachers' development of more equitable instructional practice. As reported in chapter 3, the MIST team found that teachers' views of their students' mathematical capabilities influenced whether they developed ambitious and equitable instructional practices. That is, teachers are unlikely to develop more ambitious and equitable instructional practices unless they view *all* their students as capable of engaging in rigorous mathematics.[22] When highly skilled coaches model ambitious and equitable practices with teachers' actual students, teachers have opportunities to see all their students, especially those that they perceive are currently struggling, engage in rigorous mathematics.[23]

Co-teaching

At its most basic level, co-teaching involves a teacher teaching alongside a more accomplished colleague, such as a coach.[24] In this respect, co-teaching makes sense in light of broadly accepted theories about human learning, which highlight that co-participation with someone more knowledgeable is central to learning how to enact new forms of practice.[25] Whereas modeling can provide an image of possible instruction, co-teaching is useful because it provides opportunities for teachers and coaches to engage jointly in the kinds of in-the-moment problem solving that teaching requires. At a practical level, co-teaching involves both assistance with the logistics of a lesson (i.e., passing out papers, preparing visual aids) and also verbally interjecting during a lesson at opportune moments, such as when a teacher needs to highlight, draw out, or clarify a student's thinking. The findings of several studies indicate that co-teaching can support teachers' development of new practices in the context of their classrooms work. There is also evidence that it can encourage teachers' critical analysis of ambitious and equitable instruction, especially when followed by debriefing conversations that focus on how and why the focal instructional practices influence students' learning.[26]

The Coaching Cycle

A third one-on-one activity, the coaching cycle, also satisfied the five above criteria for potentially productive activities.[27] Coaching cycles typically incorporate a pre-conference in which a teacher and coach co-plan a lesson that aims at explicitly defined learning goals. The coach then either co-teaches the lesson or observes the teacher leading instruction, and finally the teacher and coach debrief on the lesson and the students' learning. Ongoing work by Jennifer Lin Russell and colleagues at the University of Pittsburgh indicates that coaching cycles can support teachers' development of ambitious and equitable instructional practice.[28] Early findings suggest that the planning phase can support teachers in understanding the rationale for particular instructional practices, as well as how these practices can be adjusted for the teachers' particular students. There is also evidence that the observation/co-teaching and debriefing phases can support teachers in unpacking the work that they did with students.[29] In addition, the coaching cycle might support the teacher's and coach's development of a shared discourse about instruction and students' learning. One of our partner districts attempted to implement the coaching cycle as a primary means of supporting teachers. However, the coaches in this district reported that they rarely implemented the full cycle because they found it difficult to schedule all three phases with teachers.[30] This indicates that logistical issues will need to be resolved if district- and school-based coaches are to enact this promising activity with teachers on a regular basis.

Observing Instruction and Providing Feedback

An alert reader will have noticed that although coach observation and feedback is a part of the coaching cycle, we have not included it as a potentially productive coaching activity. This may be surprising given the prevalence of this activity. Indeed, when we began working with our four partner districts, coaches observing a lesson that the teacher had planned on his or her own and then providing feedback without a follow-up debriefing conversation was, by far, the most common one-on-one coaching activity. Although coaches modeled instruction and co-taught more frequently over time in response to MIST feedback, observation and feedback continued to be the most common activity.

Despite the prevalence of coach observations and feedback, there is insufficient evidence from research to classify it as potentially productive.[31] We speculate that observation and feedback might be appropriate when combined with other activities that support teachers to understand the rationale for a particular instructional practice. Furthermore, it may be appropriate when a teacher attempts to enact a particular practice on his or her own, since the feedback could help the teacher refine his or her enactments of the practice. However, coaching activities that enable teachers to learn about the rationale for new teaching practices, and to understand how to adapt and modify the practice appropriately, such as modeling, co-teaching, and the coaching cycle, seem essential.

WORKING WITH GROUPS OF TEACHERS

We also found evidence that four coaching activities conducted with groups of teachers can support their learning: (1) engaging teachers in mathematics;

(2) analyzing student work; (3) analyzing classroom video; and (4) engaging in lesson study.[32]

Engaging Teachers in Mathematics

There is evidence that attempting to solve rigorous mathematical tasks and then discussing solutions can challenge teachers' views of disciplinary norms, support their development of disciplinary knowledge that is specific to the problems of teaching, such as mathematical knowledge for teaching, and enables them to identify the big disciplinary ideas that are the focus of instruction.[33] These developments are important because, as was pointed out in chapter 3, teachers who develop ambitious and equitable practices come to view mathematics as a discipline in which solving novel tasks is valued.[34] Because many teachers were not introduced to mathematics as a problem-solving endeavor, recognizing the centrality of problem solving is an important step that challenges many teachers' notions of mathematics as a discipline.

There are also indications that engaging teachers in mathematics can support them to think more deeply about students' thinking.[35] When teachers engage in mathematics themselves, they can be given opportunities to anticipate students' strategies and to consider how they might respond. Focusing on students' thinking is an essential aspect of ambitious and equitable instruction.

Analyzing Student Work

There is growing consensus that understanding and being responsive to students' thinking is central to instruction that is ambitious and that broadens access to key mathematical ideas.[36] There is also evidence that analyzing students' written work can support teachers in learning about their students' current thinking and in using this understanding to plan upcoming lessons, rather than merely to assess whether students "got it" or not. In addition, analyzing student work can support the development of a common language for describing student understanding and of routines for eliciting and pressing on students' thinking.[37]

Analyzing Classroom Video

Group analyses of classroom video can support teachers in making sense of ambitious and equitable instructional practices. In particular, watching classroom video can support teachers in developing a vision of ambitious and equitable instructional practices, such as eliciting and pressing on student thinking, and also in identifying aspects of their own teaching they want to further develop.[38] In one study, teachers who participated in video clubs with an accomplished facilitator improved in supporting students to make their thinking public, eliciting multiple problem-solving strategies from students, and pressing students for their underlying understandings.[39]

Lesson Study

In lesson study, small groups of teachers typically work together for several months to improve particular lessons, often with the guidance of a coach.[40] Teachers first collaborate to develop a detailed lesson plan, one teacher then teaches the lesson while others observe, and finally the group analyzes the observed lesson in order to further improve the lesson plan.[41] In variations of lesson study, teachers co-teach a lesson rather than observe a demonstrated lesson.[42] Lesson study has the potential to support teachers' development of ambitious and equitable practices because it allows teachers to collaboratively investigate how to achieve a mathematical agenda by building on the students' current thinking. There is evidence that lesson study can support teachers in deepening their understanding of both students' thinking and disciplinary content. For instance, lesson study might support teachers in making sense of students' explanations and in selecting future tasks in light of students' current thinking. There is also evidence that lesson study may support the development of stronger professional communities by fostering productive norms of collaboration, mutual accountability, and shared frameworks for planning, enacting, and analyzing practice (i.e., teacher groups might use common instructional tasks and develop common lesson plans).[43]

In discussing each of the above activities, we have spoken of *potentially* productive coaching activities and of what teachers *can* rather than *will* learn as they engage in the activities. We have done so to emphasize that effectiveness of the activities in supporting teachers' learning depends crucially on coaches' expertise in enacting the activities with teachers. In the next section, we briefly summarize current research and share MIST findings on what coaches need to know and be able to do to support teachers' learning. This work has immediate implications for practice because it can inform both the hiring of coaches and the design of professional development for coaches.

ENACTING POTENTIALLY PRODUCTIVE ACTIVITIES EFFECTIVELY

The literature on instructional coaching makes it clear that it is difficult and complex professional work, requiring substantial expertise.[44] In this section, we describe five key aspects of coaching expertise: (1) content-specific pedagogical expertise; (2) productive views of students' current mathematical capabilities, (3) relationship-building skills; (4) a professional vision for coaching, and (5) group facilitation skills.

Content-specific Pedagogical Expertise

Most studies of coaching take as a baseline expectation that coaches should themselves be accomplished teachers who have a deep understanding of disciplinary content that is specific to teaching and a wide repertoire of content-specific instructional practices. There is evidence that coaches' content-specific pedagogical expertise can influence teachers' learning.[45] As we will discuss in chapter 8, the MIST team found that there was a relationship between the sophistication of coaches mathematical knowledge for teaching (MKT) and whether the MKT of the teachers supported improved.[46] This finding substantiates prior work: Coaches need content-specific pedagogical expertise.[47]

Productive Views of Students' Current Mathematical Capabilities

Our focus on instruction that is equitable as well as ambitious leads us to highlight coaches' views of students' current mathematical capabilities (VSMC) as an essential aspect of coaching expertise. As discussed in chapter 2, VSMC refers to the extent to which teachers (and coaches) see all their students as capable of participating in rigorous mathematics. Members of our team found that, even when supports for teacher learning were of high quality and teachers had a relatively sophisticated understanding of high-quality math instruction, their classroom practices did not improve unless they also saw students as capable of engaging in rigorous mathematics.[48] We therefore highlight the role of VSMC in accomplished coaching as well as in accomplished teaching. Encouragingly, we found that the coaches in our four partner districts had developed more productive views of students' mathematical capabilities than the teachers they were charged with supporting.

Trust and Relationship-Building Skills

Research on coaching consistently indicates the importance of interpersonal, trust-building skills.[49] If coaches are to co-teach, model instruction, and analyze the work of students in teachers' classrooms, teachers will need to open up their classrooms. It is therefore essential that teachers trust coaches to support them in improving their instruction rather than merely evaluating their current instruction.[50] The same is true in coaches' work with groups of teachers: In group discussions that support teachers in making sense of new instructional practices, it is important that teachers describe their current problems of practices and perhaps also share videos of their instruction.[51] Teachers frequently note that making their work public in this way can be intimidating, especially since teaching has historically been a privatized profession in which teachers have autonomy in their classrooms.[52] Thus, coaches need to build trust with teachers if they are to have access to teachers' classroom practice. In our analysis of interview data from two consecutive years in one district, coaches typically reported that they encountered fewer problems building trust when they worked fulltime at a single school or served schools for more than one year.[53] We see coaches' trust-building skills as integral to what is described in chapter 5 as coaches' ability to support a group of teachers in becoming a community of learners.[54]

A Professional Vision for Coaching

Being an accomplished teacher who has developed deep mathematical knowledge for teaching, productive views of students' mathematical capabilities, and relationship building skills is essential to coaching expertise. However, these aspects of coaching expertise tell us little about how coaches consistently engage teachers in potentially productive coaching activities, when they use these activities, with whom, and to what purpose for teachers' learning. The MIST team investigated these questions and concluded that accomplished coaches draw on what we term a *professional vision for coaching* to make these decisions.[55] It is often said that coaches need to understand processes of adult learning.[56] A professional vision for coaching specifies what coaches need to understand not just about adult learning in general, but about teachers' learning more particularly. Delineating this professional vision for coaching clarifies an aspect of coaching expertise that goes beyond being an accomplished teacher. It is useful because it can both guide the design of professional development for coaches and inform hiring decisions for instructional coaching positions.

Members of the MIST team analyzed how one accomplished mathematics coach made decisions about when to engage particular teachers in specific one-on-one coaching activities. Teacher and coach interviews over the first four years of our study indicated that this coach consistently engaged the seven mathematics teachers at her school who were in our study in potentially productive one-on-one coaching activities, such as modeling and co-teaching, and well as in observing and debriefing classroom teaching.[57]

We found that this coach drew on her sophisticated vision of high-quality mathematics instruction (VHQMI) to assess individual teachers' current instructional practices. When she observed classroom instruction, she attended to students' learning opportunities and to teachers' current practices, and located those practices on a long-term trajectory of teachers' development that culminated with ambitious and equitable instructional practices. She then identified immediate goals for individual teachers that constituted next steps in their learning, such as supporting the teacher's development of mathematical knowledge for teaching (MKT) or the teacher's understanding and enactment of specific instructional practices (e.g., launching rigorous instructional tasks so that all students can begin working productively on tasks). Next, she made decisions about the types of coaching activities in which to engage individual teachers by matching the purpose of activities with the immediate learning goals she had identified for particular teachers. For example, she explained that modeling instruction was appropriate when a teacher questioned whether her students could participate effectively in ambitious instruction organized around challenging non-routine tasks, because modeling allowed teachers to see that their students were capable to engaging in rigorous mathematics. She proposed that co-teaching, on the other hand, was appropriate when a teacher needed to experience and come to appreciate the value of a specific instructional practice in supporting all students' learning.[58] In our view, developing a professional vision for coaching is an important goal for coaches' learning that can orient the design of professional development for coaches.

Group Facilitation Skills

Consistent with prior research, our analyses of teacher collaborative groups in our partner districts, reported in chapter 6, indicate that teacher groups require expert facilitation.[59] An analysis of teacher groups in two of the districts indicates that only one-third of the groups discussed instruction in ways likely to support the teachers' development of ambitious and equitable practices.[60] This finding signals the importance of ensuring that teacher workgroups have an expert facilitator to guide their collaborative work. Coaches are well positioned to provide this facilitation.

Findings on effective group facilitation echo our findings on facilitating pull-out professional development reported in chapter 5: Skilled facilitators press teachers to make their pedagogical reasoning explicit and attend to their rationales for their pedagogical decisions. When teachers' rationales are not fully developed or articulated, it is essential that facilitators support and press them to explain their pedagogical reasoning while also preserving trust.[61] In doing so, skilled group facilitators support teachers to provide detailed descriptions and analyses of students' mathematical thinking, to relate that thinking to the instruction that students received, and to consider how instruction might be improved to support students' learning more effectively.[62] In short, skilled facilitators help teacher groups to connect their students' thinking, their instruction, and their mathematical learning goals.

Skilled facilitators also set feasible goals for teachers' learning and select activities and materials in light of those goals, much like coaches who work effectively one-on-one with teachers in their classrooms.[63] For instance, if a teacher group is going to analyze classroom video, the video should be selected because its analysis provides opportunities for the teachers to work toward particular learning goals that constitute significant steps in their long-term development. The same is true when teacher groups engage in mathematics and when they analyze student work: the tasks need to be selected with feasible learning goals for teachers in mind. These findings suggest that a professional vision for coaching might be an important aspect of coaching expertise when working with groups of teachers as well as when working with teachers one-on-one.[64] However, it is also important to note that the coach we studied who was highly skilled in one-on-one coach activities struggled to facilitate teacher groups with the same level of sophistication. This suggests that, while there are likely similarities in the capabilities required for coaching one-on-one and in groups, there might also be important differences.

Summary of Aspects of Coach Expertise

We have highlighted five important aspects of coach expertise inherent in enacting potentially productive activities effectively: content-specific pedagogical expertise, a productive view of students' current mathematical capabilities, trust and relationship-building skills, a professional vision for coaching, and group facilitation skills. Although we have discussed these aspects of coaching expertise separately for ease of exposition, we assume that they are different aspects of a mutually reinforcing set of capabilities. For example, coaches have to establish relationships grounded in trust in order to work with teachers effectively. Those relationships are likely to deepen as teachers engage in activities that they find meaningful and that support them in improving their instruction. As a second example, strong content-specific pedagogical expertise likely supports coaches' delineation of a trajectory for teachers' learning that culminates with their development of ambitious and equitable instructional practices. We anticipate that the aspects of coaching expertise we have discussed will prove to be incomplete and look forward to future research that further clarifies what coaches need to know and be able to do to support teachers in improving the quality of their classroom instruction.

THE INFLUENCE OF SCHOOL AND DISTRICT CONTEXTS ON COACHES' PRACTICES

The work that we have discussed thus far aimed to identify potentially productive coaching activities, and to tease out important aspects of coaching expertise that influence whether that potential can be realized when coaches enact the activities with teachers. Yet, substantial empirical evidence indicates that coaches do not necessarily spend the bulk of their time engaging teachers in activities likely to support the development ambitious and equitable instruction.[65] Although the primary responsibility of coaches who participated in our study was to support teachers' learning, they often spent a significant proportion of their time on other duties, such as working with achievement data, performing managerial or administrative tasks, and tutoring students.[66] This observation is consistent with the findings of prior research. For example, in one study, the participating instructional coaches spent only 23 percent of their time working with teachers on issues of instruction.[67] Similarly, Poglinco and her colleagues found that that the "overall lack of coach time was a powerful theme across both coach and teacher responses."[68] The mathematics coaches in our participating districts reported spending somewhat more time working with teachers on issues of instruction than was found in other studies of coaching. An analysis of six total years of survey data across all four participating districts indicates that the coaches spent an average of 42 percent of their time working with teachers to improve their practice.[69] We also found that coaches' time use varied significantly even within the same district, and that district-based coaches tended to spend a greater proportion of their time working with teachers than school-based coaches.[70]

We investigated this difference by focusing on one district in which five coaches had been hired by district math specialists to work across several schools, and four coaches were hired by principals to work in individual schools. The coaches hired by district math specialists were funded by the district, accountable to the district mathematics specialist, and were each responsible for supporting teachers in three to five schools. The coaches hired to work in individual schools either coached full time or spent half the day teaching and half the day coaching, and were accountable to their principal, and were funded by the school. On average, the district coaches had more sophisticated views of high-quality math instruction and deeper mathematical knowledge for teaching than their school-based counterparts.

The annual interviews that we conducted with district leaders, school leaders, and the coaches indicate that there is a tradeoff between district-based and school-based coaching models. The district-based coaches who worked in multiple schools spent an average of 84 percent of their time working directly with teachers on issues of instruction, but they struggled to build trusting relationships with teachers and principals. The high percentage of time that they spent working with teachers is not surprising given that the district leaders to whom they were accountable expected them to support teachers' implementation of the district's ambitious mathematics curricula by providing job-embedded PD and by working with teachers one-on-one in their classrooms. The difficulties that they experienced in developing trusting relationships with teachers are also understandable given that they served several schools and could often visit a particular school only once a week. Although understandable, district coaches' struggle to build productive relationships with teachers is a major concern, given that trust is essential for effective coaching.

Equally concerning, the district coaches reported that they did not work closely with principals, and communicated with them primarily through email or quick hallway conversations. These brief, opportunistic meetings are problematic as the findings of several studies indicate that coaches' relationships with principals influence whether they can work effectively with teachers on instructional issues.[71] A MIST study of teachers' advice networks, discussed in more detail in chapter 8, indicates that, in schools where coaches and teachers worked together frequently, coaches and principals met often to discuss their classroom observations and teacher collaborative meetings.[72] In contrast, coaches who did not work closely with the principal to improve mathematics instruction frequently struggled to develop trusting relationships with teachers. Principals also noted that district coaches' visits were too infrequent for them to become integral members of the school community.[73] Evidence from studies of school instructional leadership, discussed in chapter 12, indicate a further benefit of productive working relationships between coaches and principals: Principals without a strong background in mathematics education may need to rely on a district math specialist or a knowledgeable coach if they are to develop and implement instructional improvement plans capable of supporting teachers' learning.

In this same district during the same years, coaches hired by principals to work in a single school described a different set of challenges. As might be expected, coaches hired to work at an individual school were able to establish stronger relationships with teachers and the principal. However, principals often assigned these coaches duties unrelated to supporting teachers' improvement of their instruction, such as substitute teaching, tutoring individual students, identifying students for intervention initiatives, locating or creating curricula and materials for intervention initiatives, setting up computerized test preparation software, managing the student log-in process, and proctoring student assessments.[74] Consequently, school-based coaches spent an average of only 42 percent of their time working with teachers on instructional issues, as compared to 84 percent for district coaches in the same district during the same two years.[75]

It might seem reasonable to conclude that principals' expectations were the root cause of the difficulties that school-based coaches experienced in working directly with teachers on instructional issues. Some researchers have gone so far as to conclude that principals can act as a barrier to coaching and teacher leadership, while others have noted that the coaches in their studies did not believe that principals had a clear idea of how they might work with them productively.[76] However, all fourteen of the principals in this analysis explicitly noted that, first, coaches were responsible for supporting teachers' instructional improvement: They expected coaches to model instruction, lead embedded PD for groups of teachers, and observe classroom instruction and provide feedback. However, our analysis indicates that these well-intentioned expectations were undercut by the substantial pressure that principals experienced to increase students' achievement scores. Many of the coaches' additional duties were related to the demands of district benchmark and state accountability testing. In contrast, the district coaches spent less time on these types of tasks because they did not align with the district math specialist's expectations for their work.

The focal district subsequently implemented a coaching model that appeared to leverage the strengths of district coaching initiatives while reducing its weaknesses: The district hired district-based coaches who each worked in a single school, which might have allowed them to develop trusting relationships with teachers and to spend the bulk of their time working with teachers to improve their classroom practice. However, these coaches were accountable to principals rather than the district mathematics specialist and, consequently, were assigned additional tasks related to accountability testing.[77] This experience points to the potential value of a model in which district coaches work at a single school but are primarily accountable to district math specialists. This proposed model has as additional advantage beyond coaches spending most of their time working with teachers on instructional issues. As we have noted, the district coaches had greater content-specific pedagogical expertise than their school-based counterparts, perhaps because the district math specialist had greater expertise in identifying coaches with expertise in ambitious and equitable mathematics instruction than most principals. We observed similar differences in favor of coaches hired by mathematics specialists in our other three partner districts.

KEY TAKEAWAYS

The findings we have discussed clarify the types of coaching activities that have the potential to support teachers' development of ambitious and equitable instructional practices, aspects of coaching expertise inherent in selecting potentially productive activities and implementing them effectively, and aspects of school and district contexts that influence the time coaches actually spend working with teachers' on improving instruction. These findings can inform the design of initiatives that realize the promise of coaching by ensuring that coaches are indeed more accomplished colleagues who provide ongoing support for teachers' development of ambitious and equitable instructional practices. It is in this spirit that we offer the following takeaways.

Coaching models in which coaches are accountable to the district mathematics specialist and serve one or two schools appear promising. The findings of a number of studies consistently indicate that coaches are frequently assigned a range of additional tasks unrelated to supporting teachers' development of ambitious and equitable forms of instruction. This appears to be the case for coaches who are primarily accountable to principals, but less so for district coaches who are accountable to district mathematics specialists. However, we also found that district-based coaches who worked in several schools found it difficult to establish relationships grounded in trust with teachers and to collaborate productively with principals. Taken together, these findings suggest a district model in which coaches work in one, or perhaps two, schools, and are primarily accountable to the district mathematics specialist. In this model, coaches should be able to participate in the school community, develop strong relationships with teachers, and communicate regularly with principals. Furthermore, they should be able to spend the bulk of their time working with groups of teachers during teacher collaborative time and with teachers one-on-one in their classrooms, thereby providing teachers with more consistent, ongoing support.

Novice coaches should already have developed ambitious and equitable instructional practices, and should have productive views of students' current mathematical capabilities. The findings we have discussed go some way toward specifying the capabilities, beyond being accomplished teachers, that coaches need to develop to support teachers' learning effectively. These capabilities are extensive and include productive views of students' mathematical capabilities, relationship-building skills, a professional vision for coaching, and the ability to facilitate teachers' collaborative work. Developing these capabilities requires substantial learning and sustained support.[78] Pragmatically, it therefore appears essential that teachers who are recruited for coaching positions have already developed ambitious and equitable instructional practices and productive views of students' current capabilities. If coaches do not already have strong visions of high-quality mathematics instruction and productive views of students' mathematical capabilities, the learning curve may simply be too steep even if a district can provide substantial support for coaches.

Although this recommendation might delimit the number of qualified candidates for coaching positions, the experience of one of our partner districts indicates that districts are better off with fewer coaches who are relatively have developed accomplished instructional practice, sophisticated views of high-quality mathematics instruction, and productive views of students' mathematical capabilities than with a large number of coaches who are still struggling to develop these aspects of ambitious and equitable teaching. When we began working with the district, every middle school had a mathematics coach. However, our analyses indicate that the number of accomplished mathematics teachers in the district was relatively small and that the instructional practices of the coaches were not, on average, more advanced than the teachers they served. Based on our feedback, the district subsequently switched to a district coaching model and hired fewer mathematics coaches who were already accomplished teachers. This enabled the district, in partnership with MIST researchers, to provide PD for the coaches that built on their pedagogical strengths. The annual interviews conducted with the coaches and with teachers indicate that these coaches engaged teachers in potentially productive coaching activities far more frequently than had the prior school-based coaches.

Provide ongoing professional development for coaches. As we have indicated, accomplished teachers who become instructional coaches require ongoing professional development if they are to support teachers' development of ambitious and equitable instructional practices. The various aspects of coaching expertise that we have discussed constitute goals for coaches' learning and can thus orient the design of professional development. As the literature on professional development for coaches in extremely thin, we suggest extrapolating from the principles of teacher professional development discussed in chapters 4 and 5. These principles indicate that professional development for coaches should be sustained over time and should focus on problems of practice that coaches encounter in their daily work. It should also involve opportunities for coaches to engage in investigations and enactments of coaching practice with mathematics specialists or with more accomplished coaches.[79] Thus, coaches might analyze video of their own and others' coaching, observe a district math specialist or more accomplished coaches modeling particular coaching practices, and co-coach with a specialist or accomplished coach.

Coaches should prioritize leading teachers' collaborative time when the participating teachers do not have the expertise to take on this leadership role. As reported in chapter 6, in the absence expert facilitation, teachers' work in collaborative time was rarely deep enough to support their development of ambitious and equitable practices. A primary reason for this was that leading teacher collaborative meetings involves selecting potentially productive activities and implementing them effectively with teachers. However, it was frequently the case that none of the participating teachers had sufficient expertise in ambitious and equitable instruction to provide this type of leadership. Indeed, in three of our four districts, there were, overall, relatively few accomplished teachers. When this is the case, we suggest that coaches prioritize leading teachers' collaborative time and work with teachers one-on-one in their classrooms as time allows. One benefit of this arrangement is that it might enable coaches to build stronger relationships with teachers. As we have noted, such relationships are essential as they enable coaches to have access to teachers' classrooms. In turn, this access enables coaches to assess teachers' current practices, determine short- and long-term learning goals for teachers' learning, and better support teachers both during teacher collaborative time and when working with them one-on-one.

Coaches can play a key role in supporting coherence across the various facets of the teacher learning subsystem. As described in chapter 4, teachers' professional learning is best supported when various supports—including pull-out PD, school-based teacher collaborative time, and one-on-one coaching—are coordinated. Coaches are well positioned to support this coordination. For example, coaches might play a leadership role in pull-out PD sessions that focus on particular instructional practices. They could then either lead or participate in teachers' collaborative meetings that focus on the same instructional practices while keeping teachers' school contexts in view. For example, teachers might analyze student work that was completed after the group co-planned enactments of the practices. Further, coaches might model or support the teachers in enacting the practices in their classroom. In this way, teachers would have ongoing opportunities to investigate and enact ambitious and equitable instructional practices while receiving support from a more accomplished colleague *in the context of their daily work*, as suggested in the literature on high-quality professional learning. Coaches would then play a critical role in ensuring that the various supports for teachers' professional learning are coherent.

DIRECTIONS FOR FUTURE RESEARCH

Although research on instructional coaching has made considerable progress in the last few years, there are still significant gaps. A number of studies have investigated whether coaching initiatives support teachers' changes in practice or improve students' performance. While findings indicate that coaching can support teachers to improve their instruction, several studies also document the challenges to implementing coaching initiatives. In our view, there is a need for two lines of research, one that focuses on coaching practices and their influence on teachers' learning, and a second that focus on supporting coaches' learning.

There are several potentially fruitful avenues for the first line of work, coaching practices that influence teachers' learning. Investigations of one-on-one coaching might investigate accomplished coaches' decision-making about how and when to focus on particular aspects of a teacher's instruction (i.e., coaches' professional visions), and to what effect; and how they sequence their work with teachers to support their development over the long term. Another useful direction for research involves coach feedback: We need to know more about the types of feedback accomplished coaches give to teachers, and to what effect.

Although more studies have focused on coaches working with groups of teachers than on coaches' one-on-one work, more work is needed. For example, only a small proportion of these studies have systematically examined the facilitation of teacher groups. Additional research is needed on how accomplished coaches facilitate potentially productive group activities, particularly how they press and build on teachers' contributions and how they adjust their responses in light of teachers' current practices and their goals for teachers' learning. Studies are also needed to clarify how coaches can support teachers' development of more productive views of students' current mathematical capabilities both when working with groups of teachers and when working with teachers one-on-one in their classrooms.[80]

The second proposed line of research investigates designs for supporting coaches' learning. Rather than merely assessing whether particular designs work or not, there is a pressing need for studies that document the process of coaches' learning and the means by which it is supported, so that the designs can be improved and adapted to a range of different school and district contexts. A primary goal of this work should be to identify trajectories in coaches' learning. This work might, for example, investigate how coaches can be supported to develop a professional vision for coaching, such as assessing teachers' current practices, and identifying long and short-term goals for teachers' learning by situating their current practices on a trajectory for teachers' learning that aims at ambitious and equitable instructional practices. We look forward to further advances that build on and refine the findings and takeaways we have discussed.

EIGHT

Teachers' Advice Networks

ANNE GARRISON WILHELM, PAUL COBB,
KENNETH FRANK, AND I-CHIEN CHEN[1]

TEACHERS' COLLEGIAL INTERACTIONS with their peers can be an important support for improving instruction.[2] Studies have found that such interactions can facilitate teachers' adoption of new technology and the development of their knowledge and instructional practices.[3] This led us to investigate whether this was the case for mathematics teachers' development of ambitious and equitable instructional practices.

Prior investigations of the influence of teachers' instructional advice seeking have also demonstrated the important role that teachers' expertise plays in whether they learn through their interactions with colleagues.[4] In particular, it is not just interactions with colleagues that drive teachers' learning. Instead, it is important to consider the expertise of those colleagues and how they can support the learning of teachers with less expertise. We were particularly interested in how interacting with colleagues with greater expertise contributed to improvements in the quality of instructional practice (assessed by the Instructional Quality Assessment) and the depth of teachers' mathematical knowledge for teaching.

In order to understand this, we first sought to understand teachers' advice seeking by asking all of the mathematics teachers in our study schools (approximately 300 teachers) to take a survey every spring (starting in the second year of the study). Teachers responded to the survey by listing the names and roles of up to ten people they go to for advice or information about teaching mathematics (e.g., 8th grade math teacher, mathematics coach). We also asked them about the frequency of those interactions (e.g., daily, weekly, monthly).

Survey information about advice seeking can be used to create a diagram (called a "sociogram") of that teacher's advice network. Figure 8.1 shows an example of a sociogram of a single teacher's advice network. The focal teacher, Alice, is represented by the circle at the center of the network. The advice-seeking interactions that Alice reported are represented by directed arrows to three other individuals (Bob, Carol, and Dana). In this case, the arrows point away from Alice because she indicated that she seeks advice from those individuals. Bob and Carol are other mathematics teachers at Alice's school, and Dana is the school mathematics coach. In the sociograms, we use circles to represent teachers and triangles to represent coaches. In network speak, the circles and triangles are called "nodes." The last piece of survey information represented in the sociogram of Alice's advice network is the frequency with which she interacts with her colleagues. She reported that she seeks advice from Bob daily and Carol and Dana monthly, represented by a slightly thicker arrow (and larger arrowhead) toward Bob than those towards Carol and Dana.

Although individual teachers' advice networks are helpful for understanding their interactions, individual networks can be combined into a

FIGURE 8.1 Individual teacher advice network

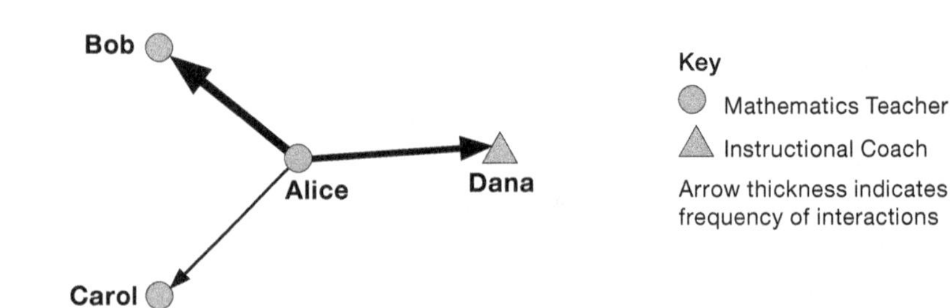

mathematics department teacher advice network that provides information about the nature of collaboration across the department. Figure 8.2 shows the mathematics department teacher advice network for Alice's school. In this example, we depict a mathematics department with eight mathematics teachers (including Alice, Bob, and Carol), all of who completed the network survey. When combined into a department network, the arrows have the potential to be bi-directional, representing a mutually reported advice-seeking interaction. For example, Alice reported that she seeks advice from Bob, but Bob also reported that he seeks advice from Alice, so in the department teacher advice network, the arrow between them is bi-directional. Several other features of advice networks emerge when individual networks are combined. For example, Erin is not connected in the network because she does not seek advice from anyone at her school, and none of the people at her school reported that they seek advice from her. Dana, the mathematics coach, seeks advice only from the school principal (who is not depicted in the mathematics department network), but six of the eight teachers seek advice from her.

FIGURE 8.2 Mathematics department teacher advice network

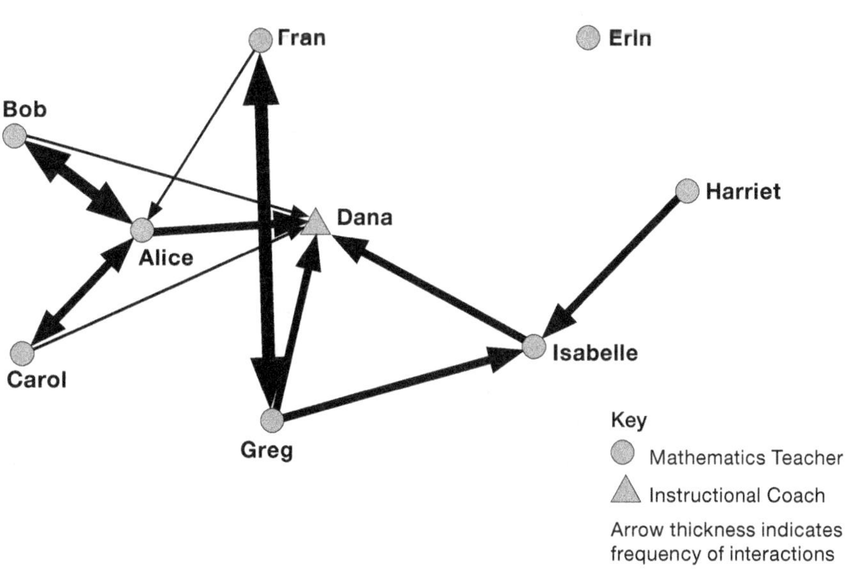

By combining the teacher network survey data with other data from our study (e.g., quality of instruction as assessed by the IQA or the depth of mathematical knowledge for teaching (MKT), grade level information), we were able to understand more about the formation of teachers' advice networks. For example, in figure 8.3, we have added information about teachers' and the coach's mathematical knowledge for teaching (MKT) to the sociogram. This is represented by the size of their nodes (i.e., circle or triangle). If they have a larger node, then they have relatively more developed MKT. For example, in this school, the instructional coach, Dana, is more expert with respect to her MKT, which is represented by a large triangle. Erin is the teacher with the most developed MKT, depicted by the largest circle, possibly contributing to her lack of advice seeking. In other words, it could be that she is aware of the depth of her mathematical knowledge for teaching expertise and does not see a reason to seek advice from colleagues whose MKT is less developed. For this example, we have chosen to focus on MKT as a measure of expertise because it is a widely used construct. Other measures (e.g., views of students' mathematical capabilities (VSMC), instructional practices (IQA), student achievement averages, years of experience teaching) could be added or used instead depending on the issues of interest. In figure 8.3, we have also added information about the grade level of the teachers by varying in the shading of the nodes, with white representing sixth grade teacher, dark grey representing seventh grade teacher, and black representing eighth grade teacher. We use a light grey triangle to represent the fact that the mathematics coach was not teaching sixth, seventh, or eighth grade at the time. By adding grade level information, it becomes clear that teachers do tend to interact more with their colleagues at the same grade level.

Teacher networks differ from other supports for teachers' improvement of their instructional practices because they are not part of districts' explicit designs for instructional improvement.[5] However, given the potential of teachers' advice seeking interactions to support their learning, a number of studies have investigated how different factors influence teachers' decisions about which colleagues to turn to for advice. These studies suggest that there are several types of factors that influence teachers' selection decisions. Further, recent research suggests that school, district, and national policy can influence teachers' decisions about advice seeking.[6] Thus, although teacher networks cannot be directly mandated, school and district leaders can

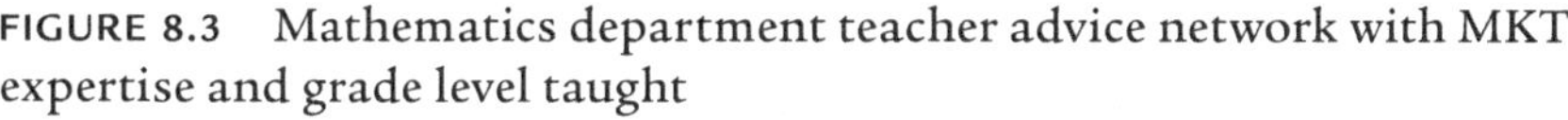

FIGURE 8.3 Mathematics department teacher advice network with MKT expertise and grade level taught

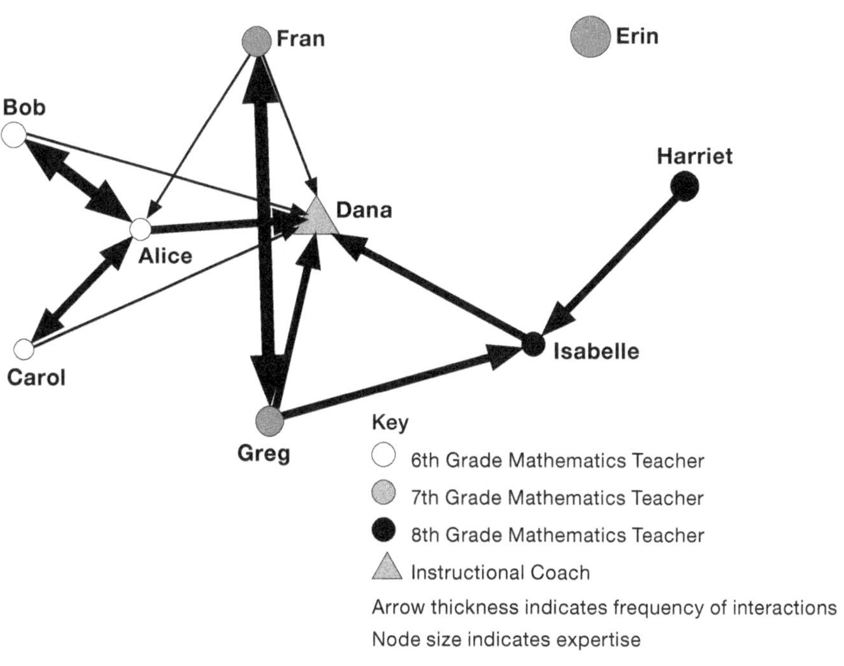

influence them, thereby influencing instructional improvement for good or ill. Our primary goal for this chapter is to underscore the importance of considering teachers' social networks when making decisions in schools.

In the MIST project, we were particularly interested in understanding more about whether and how teachers' advice networks matter for supporting the development of ambitious and equitable instructional practices, and how school and district leaders can foster productive network interactions. In what follows, we first discuss our findings about the influence of social network interactions on teachers' knowledge and practice, and then our findings about how other elements of a coherent instructional system influence teachers' decisions about who to turn to for advice. These findings are important given the accumulating evidence that teachers' network interactions can support their learning. Although rarely discussed, teachers' advice networks can support or undermine school and district goals for instructional improvement. We conclude by outlining a set of takeaways for school and district leaders and then by identifying key areas for future research on teachers' networks that is specific to the development of ambitious and equitable instructional practices.

INTERACTIONS WITH COLLEAGUES SUPPORT TEACHER LEARNING OF AMBITIOUS PRACTICES

There is mounting evidence that teachers' advice-seeking interactions can support instructional improvement.[7] We built on this prior work while focusing specifically on the development of ambitious and equitable instructional practice.[8] We examined video recordings of teachers' classroom instruction and assessments of their mathematical knowledge for teaching in the 2008–9 and 2009–10 school years to understand whether advice seeking supported improvements in their MKT and the quality of their instructional practice. We found that teachers' interactions with colleagues who had developed more ambitious and equitable instructional practices supported improvements in teachers' own instructional practices. However, we found that teachers did not necessarily develop more sophisticated mathematical knowledge for teaching (MKT) through their advice-seeking interactions. Specifically, advice-seeking interactions with more expert colleagues supported improvements in teachers' MKT only in schools in which the coach's MKT was relatively sophisticated. The coach in the illustration in figure 8.3 had developed relatively sophisticated MKT, indicating this could have been a school where teachers' MKT improved though their advice-seeking interactions.

One possible explanation for this finding concerns coaches' influence on how teachers worked together. Cynthia Coburn and Jennifer Lin Russell found that coaches' interactions with teachers impact how they interact with each other.[9] They compared two school districts and saw differences in how the mathematics coaches worked with teachers, and traced the consequences for how the teachers worked with each another. For example, in one district, coaches and teachers interacted in "high-depth" ways that focused on analyzing mathematics problems and student solution methods. In the other district, coaches and teachers spent more time discussing the logic of the curriculum and how to lead particular instructional activities in the classroom, which Coburn and Russell classified as "low-depth" because of their lower potential to support the development of ambitious and equitable practices. With regard to our findings about the influence of the coach, one explanation is that coaches who had developed sophisticated MKT were more likely to press teachers on content-specific issues (e.g., what are the big mathematical ideas in this lesson?), and that this influenced types of interactions between teachers in the school. Therefore, our findings around the influence of coaches with more sophisticated MKT on teachers' informal interactions suggests that both *who* teachers interact with and *what* they talk about both matter for the development of knowledge and practice.

In chapter 3, we reported that there is a link between teachers' views of students' mathematical capabilities (VSMC) and their enactment of ambitious and equitable instruction. We would have liked to explore the influence of teachers' interactions with their colleagues on teachers' VSMC. Unfortunately, too much teacher VSMC data were missing to investigate this question. We subsequently revised our interview procedures for VSMC and plan to explore the influence of teachers' advice seeking interaction on their views of students' mathematical capabilities in future analyses.

SCHOOL AND DISTRICT LEADER DECISIONS INFLUENCE TEACHERS' ADVICE NETWORKS

In addition to demonstrating that teachers' advice seeking interactions influence the development of their knowledge and instructional practice, we found that several other elements of a coherent instructional system can influence individual teachers' decisions to turn to a colleague for advice. In particular, we clarify how coaching, teacher collaborative time (TCT), and school leadership all influence mathematics teachers' decisions about advice seeking.

Mathematics Coaching

First, in examining MIST study data from the 2008–9 school year through the 2010–11, we found that teachers tended to seek out instructional coaches for advice.[10] The school example shown in figure 8.3 illustrates such a case as six of the eight teachers indicated that they went to Dana, the coach, for advice about teaching mathematics. On average, in our sample, a teacher was approximately 25 percent more likely to seek advice from a coach than from another teacher. However, we also found that there is quite a bit of variation in the extent to which coaches are central sources of instructional advice in their schools.[11] In other words, while on the whole, coaches are more likely to be sought out for advice, a number of factors influence whether teachers seek out advice from individual coaches. As we discuss later, in examining data from schools in District B in the 2009–10 school year, we found that the extent to which the coach and the principal coordinate their work with teachers is consequential for whether the teachers turn to the coach for instructional advice, and thus whether the coach is central in the department level advice network. As described in chapter 7, in schools where the coach was central, the principal and coach coordinated their work in several key ways including: 1) both attended teacher collaborative meeting, with the coach serving as the facilitator and the principal serving as an observer; 2) both spent time in classrooms in order to understand teachers' needs; and 3) they met regularly to discuss supports for teachers to improve their instruction.

Teacher Collaborative Meetings

Teacher collaborative meetings can have an impact on teachers' advice seeking but it is the quality of meetings rather than the quantity of meetings that matter for the development of ambitious practices. First, several MIST studies and other research provide strong evidence that time in collaborative meetings does not necessarily influence teachers' practice. For example, one MIST analysis investigated the influence of several factors, including several different measures of teachers' expertise (e.g., IQA, MKT, VHQMI, and student achievement gains), the grade level they taught, and their participation in collaborative meetings and found that time spent in collaborative meetings was not influential.[12]

However, the findings of a follow-up analysis indicate that the quality of the conversations in teacher collaborative meetings influences teachers' advice seeking outside TCT.[13] As discussed in greater detail in chapter 6, in high-quality TCT, teachers analyzed instruction in ways that were likely to deepen their knowledge and improve their classroom practice. For example, teachers might analyze student work to investigate students' understanding of a particular mathematical concept, and then use what they learned to plan for future instruction.[14] In contrast, in lower quality TCT, teachers tended to talk about logistics or to share instructional strategies without digging into why the strategies might support students' learning.[15]

We found that teachers were more likely to seek advice from one another outside of meetings if TCT was of high quality, and those advice-seeking relationships tended to last. In particular, a follow-up analysis suggests that the influence of productive TCTs on teachers' advice networks holds up for at least a year. It is likely that teachers learned about their colleagues' expertise and began to build trust in these meetings, which led them to turn to each other for advice outside of meetings. In sum, our findings demonstrate that while time in TCT is necessary but not sufficient, the quality of TCT discussions influences both teachers' learning and their decisions about advice seeking.[16] Therefore, investment in productive TCTs not only produces greater opportunities for learning within the meetings, as discussed in greater detail in chapter 6, but it can also lead to more productive teacher collaboration outside of those meetings, another important site for teacher learning.

School Leaders

In our districts, teachers rarely went to school leaders for advice about teaching mathematics.[17] However, there is considerable evidence that their decisions influence teachers' advice seeking. First, described above and discussed in chapter 7, we found that in schools where coaches were central sources of advice for teachers, the coach and principal coordinated their efforts to support teachers.[18] In particular, in the schools in one district in which the coaches were central sources of advice for teachers, the coach and principal jointly set a vision for and attended teacher collaborative meetings, both visited classrooms, and they met frequently to support each other's work and to discuss appropriate supports for teachers.

Second, school leader decisions about assigning people to particular roles matter. For example, consistent with a number of other studies, we found that when people are assigned to formal roles (e.g., instructional coach, department chair), the chances that they will be sought out for instructional advice increases.[19] Further, grade level teaching assignments influence teachers' advice seeking as teachers are more likely to seek advice from colleagues who teach in the same grade level.[20] The decision about where to place a new teacher should therefore be more than a matter of filling a vacancy at a particular grade level, and should take into account whether there are colleagues at that grade level who can effectively support a new teacher.

Finally, school leaders' expectations for teachers (e.g., improving the quality of their instructional practice, improving students' performance on procedural assessments) and how they communicate those expectations can impact on teachers' decisions about advice seeking. For example, we found that the mathematics department advice network in one school fragmented over three years when the emphasis that the principal set for teacher collaborative meetings shifted from a focus on developing ambitious and equitable instructional practices to a focus on improving scores on procedurally oriented assessments.[21] In particular, over time, as the principal increased his press in meetings for improvements in students' scores on procedurally oriented district benchmark and state assessments, there was less talk about mathematical concepts in meetings, an overall decrease in teacher advice seeking, and teachers tended to turn to colleagues with the highest student achievement scores rather than the most sophisticated instructional practices. One hypothesis is that this occurred because the focus on test scores left little for teachers to discuss instructionally, and thus there was little reason for teachers to seek advice from colleagues. In addition, a focus on test scores may have led the teachers to perceive that they were in competition and were therefore more hesitant to collaborate.

In another analysis, we found across our four partner districts that teachers tended to seek advice from colleagues who had improved their students' test scores rather than those who had developed other forms of teacher expertise (i.e., ambitious and equitable instructional practice, deep MKT, sophisticated instructional vision).[22] This is likely because expertise in improving test scores was valued by principals in the study schools and was more visible to colleagues than some of the other forms of expertise (e.g., accomplished instructional practice). For example, it could be that principals publicly discussed certain teachers' effectiveness in improving students' test scores, which signaled to teachers that the principal particularly values this form of expertise. This finding suggests that if principals want teachers to seek advice from teachers who have developed relatively sophisticated instructional practices, then they should indicate that this form of expertise is valued and help teachers learn about their colleagues' instructional practices (e.g., by observing one another's instruction).

We found that teachers rarely considered colleagues' MKT when determining who to turn to for advice and hypothesize that they may not have known much about the depth of their colleagues' MKT. The illustrative department network shown in figure 8.3 is representative as Erin had developed the deepest MKT yet her colleagues did not seek advice from her. If we had shown student achievement averages in the figure rather than MKT, we might see a different relationship between advice seeking and this form of expertise. For example, it might be that Erin's students did not perform particularly well on procedurally oriented standardized assessments for some reason and was therefore not regarded as having relevant expertise.

Summary

The findings we have reported illustrate how both elements of a coherent instructional system and school leaders' decisions influence teachers' advice seeking. In particular, we found that coaching, teacher collaborative meetings, and school leaders influence teachers' advice seeking. Other studies have demonstrated that pull-out professional development and curriculum materials can also influence teachers' advice seeking.[23] Therefore, there is clear evidence that while social networks cannot be directly designed, school leaders can support and influence the development of productive teacher advice networks.

TAKEAWAYS FOR MAXIMIZING THE POTENTIAL OF TEACHERS' ADVICE NETWORKS

We offer several key takeaways from our research on teachers' advice networks and elaborate on how they connect to other elements of the coherent instructional system. First, **the distribution of expertise matters**. In particular, teachers need access to colleagues with expertise in order to learn from their interactions. Further, teachers need to know about their colleagues' expertise if they are to make informed decisions about advice seeking. However, it is also important to consider the different forms of expertise. Some forms of expertise are more commonly known by school staff (e.g., expertise in improving student achievement) whereas other forms of expertise only become known through interactions or experience (e.g., mathematical knowledge for teaching, views of students' current mathematical capabilities, sophistication of instructional practice). For example, in the illustrative network in figure 8.3 above, it could be that the teachers are unaware of the depth of their colleagues' mathematical knowledge for teaching, and it might therefore not have factored into their decisions about advice seeking. However, school leaders' actions can make this and other less visible forms of expertise more visible. For example, asking teachers to observe one another teaching can make the practice of accomplished colleagues more visible. In addition, a coach who teaches a teacher's students can make the students' current capabilities more visible. If teachers are made aware of particular forms of expertise and school leaders communicate that those forms of expertise are valued, they may be more likely to use that information to make decisions about advice seeking, and then be more likely to learn through their advice-seeking interactions.

Second, **the content and purposes of interactions matter**. In other words, what teachers talk about when they seek advice from colleagues and why they talk about it matter for their learning. We found that coaching, teacher collaborative time, and principals' expectations can all have an impact on teachers' interactions. By influencing the content and purposes of interactions, these elements of the instructional system have an impact on teachers' advice seeking and thus on their learning. Therefore, the supports that school leaders provide and the expectations they communicate for coaching and teacher collaborative time influence their effectiveness, and also impact teachers' one-on-one interactions with colleagues in other settings.

Third, **routine decisions should be made with social networks in mind**. For example, considering social networks when making staffing decisions means taking account of the current distribution of valued forms of expertise. In particular, school leaders should place novice teachers in grade levels or teams that have developed sufficient relevant expertise. For example, the wisdom of replacing a retiring third-grade teacher with a novice teacher depends on whether the other third-grade teachers have sufficient expertise to support that teacher's development of ambitious and equitable practice. If they do not, it is perhaps better to move an accomplished veteran teacher into third grade to support the current third-grade teachers' learning, and place the novice teacher at a grade level where the teachers have sufficient expertise. Consistent with the network shown in figure 8.3, research suggests that teachers tend to seek advice from their grade level colleagues so it is important that the teachers at that grade level have developed sufficient expertise to support the novice teacher.

A practical question that arises is how school leaders can collect information about teachers' advice networks that can inform their decision making. As we indicated, we used a survey to collect information about who teachers interacted with and the frequency of those interactions. The survey was fairly easy to administer and did not take much time for teachers to complete. While the sociogram shown in figure 8.3, created using NetDraw, provides a way to visualize a network, it is also feasible to organize this survey data in a spreadsheet given the relatively small number of teachers and school leaders' (and coaches') familiarity with them.[24] School leaders could quickly look at the spreadsheet to see who is listed frequently as a source of instructional advice, and whether particular individuals are isolated within the network for one reason or another. By combining this information with information about teachers' expertise school leaders can understand how teachers' advice networks are functioning and make informed decisions based on this knowledge.

One final takeaway concerns the **interconnections between the elements of the instructional system**. While teachers' advice networks cannot be mandated or directly designed, we found that coaches, teacher collaborative time, and school leaders influence teachers' social networks in a number of ways. Other research suggests that additional aspects of the instructional system (e.g., pull-out professional development, curriculum materials) can also influence both whether teachers turn to others for advice and the content of those interactions.[25] Further, this influence can be either positive or negative in nature. On the one hand, the teacher learning that is supported by productive networks comes for free with well-designed supports for teachers' learning (e.g., high-depth conversations in TCT, coach-principal coordination of teacher support). On the other hand, the unintended consequences of some decisions can undermine the potential of advice networks to support teachers' learning. It is therefore important when designing and implementing strategies for supporting teachers' development of ambitious and equitable instructional practices to carefully consider social networks together with the other elements of the instructional system.

FUTURE RESEARCH

In conclusion, we identify two key areas for future research regarding teachers' advice networks and the development of ambitious and equitable practices at scale. The first key area for future research requires just a slight shift in orientation. As a field, **we need to know more about the role of teacher**

networks in supporting their development of classroom practices that are equitable as well as ambitious. It is likely that teachers' networks are consequential for attending to issues of equity in students' learning, but we need to know more about exactly how teacher network interactions can support the development of more productive views of students' current mathematical capabilities and more equitable instructional practices.

The second key area for future research pertains to **the role of teacher networks in the development of a shared vision of high quality instructional**. Our partner districts each pursued an ambitious and equitable vision of high-quality instruction. However, while necessary, such a district vision is not sufficient unless it is shared by teachers, coaches, and principals across the district. Given the existing research on teachers' tendencies to adapt their conceptions or practices to match those within their networks (i.e., teachers adopting the conceptions of their colleagues through informal interactions), it is likely that social networks play a role in the development of a shared instructional vision, but this hypothesis is yet to be investigated.[26]

NINE

Instructional Materials as Tools for Instructional Improvement

ERIN HENRICK, MOLLIE APPELGATE,
AND MAHTAB NAZEMI

IN THIS CHAPTER, we focus on the role of instructional materials as tools for supporting instructional improvement. *Instructional materials* include both the adopted textbook and district-developed resources such as curriculum frameworks and pacing guides. Unsurprisingly, mathematics instructional materials are not all created equal, but vary significantly in terms of the rigor of their learning goals and opportunities they provide for students to attain those goals.[1] High-quality instructional materials provide greater opportunities to attain rigorous goals and include high-cognitive demand tasks that are sequenced to support students' development of significant mathematical ideas.

We know from prior research that teachers' use of high-quality instructional materials can also support their learning.[2] Instructional materials designed to support teachers' development of ambitious and equitable instructional practices are therefore an integral element of a coherent instructional system. In what follows, we first describe the instructional materials that our four partner districts used and then we clarify the issues we investigated before sharing our findings. We conclude with takeaways for district leaders seeking to support teachers to use high-quality instructional materials effectively, and with recommendations for future research in this area.

INSTRUCTIONAL MATERIALS USED IN OUR PARTNER DISTRICTS

As described in prior chapters, all four MIST districts aimed to support middle-grades students' attainment of rigorous goals in mathematics that included both conceptual understanding as well as procedural fluency. Three of the districts adopted *Connected Mathematics Project 2* (CMP2) as their primary text.[3] CMP2 takes an inquiry-oriented approach and was well aligned with the districts' learning goals. The fourth district had an adopted textbook that aimed at less ambitious learning goals as its primary instructional resource. However, the district also provided CMP2 to schools as a secondary resource, and the district mathematics department developed supplemental instructional materials that included high-cognitive demand tasks.

One of the three districts that adopted CMP2 had used the previous version of these materials, the CMP1 curriculum. The other two districts adopted CMP2 at the beginning of our partnerships with them and provided teachers with ongoing professional development on using these materials effectively. Two of these three districts had moved away from requiring that teachers use CMP2 by the end of our partnerships with them. One of these districts revised the textbook adoption policy following a change in senior district leadership and allowed individual schools to choose instructional materials. In the second district, the textbook adoption process resulted in the selection of a more traditional textbook series, to the dismay of the district mathematics department.

In addition to textbooks, two of the districts developed pacing guides intended to align instruction with state standards by providing guidance on *what* to teach and *when* to teach it. The other two districts developed more elaborate curriculum frameworks that not only provided guidance on what should be taught and when it should be taught, but also on *how* to teach it. Examples of the guidance provided in these frameworks include descriptions of the big mathematical ideas that were the focus of instructional units and individual lessons, recommended instructional strategies, descriptions of typical student misconceptions, additional tasks to support differentiation, and formative assessment items and strategies.

HOW WE STUDIED TEACHERS' USE AND VIEWS OF INSTRUCTIONAL MATERIALS

The data we collected each year to assess the quality of instruction included video-recordings of two lessons in each teacher's classroom together with documentation of the tasks the teacher used and the source of the tasks (e.g., CMP2, other textbooks, or websites). We used the Instructional Quality Assessment (IQA) rubrics to code tasks in terms of their level of rigor and distinguished tasks that were procedural in nature from those were more challenging and gave students' opportunities to develop conceptual understanding and problem-solving capabilities (see chapter 2).[4] We also used the IQA rubrics to code the video-recordings of classroom lessons in order to assess whether teachers maintained the level of cognitive demand of tasks during instruction.

We assessed teachers' views of the instructional materials adopted by their school or district by asking in the annual interviews about the appropriateness of the materials for all of their students. In addition, the survey that teachers completed each year included a question about whether or not the materials were consistent with their personal views about effective mathematics teaching.

FINDINGS

Influence of Instructional Materials on the Cognitive Demand of the Tasks Selected by Teachers

As described in earlier chapters, ambitious and equitable teaching involves the use of high-cognitive demand tasks designed to support students' attainment of rigorous learning goals. Consistent with prior research, we found that the level of cognitive demand of the tasks that teachers used was related to students' performance on standardized state assessments.[5] Specifically, in an analysis of data collected in all four districts in years one through four, when the state student assessments were procedural in nature, indicated that the difference in student value added scores in classrooms in which high-cognitive demand and low-cognitive demand tasks were used was equivalent to an additional 22–30 additional days of mathematics instruction. Moreover, more than half the teachers reduced the cognitive demand of tasks during instruction, typically by demonstrating specific steps to use to solve the tasks (see chapter 3).[6] Taken together, these findings indicate that there might be benefits in teachers using rigorous, cognitively demanding tasks even if they proceduralize the tasks by showing how students how to them. In our view, this finding is significant given that the development of ambitious and equitable instructional practice is demanding and requires support for an extended period of time (see chapter 3).

In light of the above finding, we looked at the cognitive demand of the tasks that teachers used by district and found that, on average, the teachers in the three districts that had adopted CMP2 used significantly more rigorous tasks than those in the fourth district where high-quality materials were supplemental.[7] We also looked at the district that moved away from the district-wide use of CMP2 following a change in senior leadership and found the cognitive demand of the tasks that teachers used decreased after this change was made. In the last year that CMP2 was used district wide, 85 percent of the participating teachers selected high-cognitive demand tasks. However, two years later, less than 70 percent of teachers used high-cognitive demand tasks as the basis for their instruction. These findings emphasize the importance both of making high-quality instructional materials available to teachers and of supporting and pressing teachers to use them.

Teachers' Views of High-Quality Instructional Materials

We should clarify that nearly one-third of the lessons video recorded in the districts that had adopted CMP2 were not based on CMP2 tasks. If the teachers were substituting other high-cognitive demand tasks for CMP2, this might not be a problem. However, the above findings indicate that non-CMP2 tasks were often of lower cognitive demand. We therefore looked at teachers' responses to the interview and survey questions to understand this issue and found that there were two primary reasons why some teachers were not selecting CMP2 tasks. First, many teachers' views of high-quality math instruction did not align with the launch-explore-summarize structure of CMP2's lessons and with the recommended instructional strategies. An analysis of four years of survey data in the three districts that had adopted CMP2 as the primary text indicated that almost one-third of teachers' views of high quality instruction did not align with or was only minimally aligned with CMP2. A second, follow-up analysis of one year of interview data in two of the districts that had adopted CMP2, 79 percent of the teachers said that CMP2 was not appropriate for all their students. When asked why, teachers most frequently reported that some students lacked pre-requisite mathematical or reading skills, and that the materials did not provide enough step-by-step examples and practice tasks.

Taken together, our findings indicate that teachers' visions of high-quality instruction and their views of the appropriateness of rigorous materials for their students can be consequential for the quality of their instruction and thus for student learning. These findings are consistent with findings reported in chapter 3 regarding how teachers' visions of high quality instruction and their views of their students' current mathematical capabilities impact whether they capitalize on available supports to improve their instruction.

Teachers' Use of Pacing Guides and Curriculum Frameworks

As described above, two of our partner districts developed pacing guides that indicated *what* to teach and *when* to teach it, and the other two districts developed more elaborate curriculum frameworks that also provided guidance about *how* to teach. In the annual interviews, we asked teachers about whether and how they used these instructional resources. We found that most teachers in the two districts that developed pacing guides used them as intended after only one or two professional development sessions. However, we found that most teachers in the other two districts ignored the guidance provided in the curriculum frameworks about *how* to teach and used them primarily as pacing guides even after attending professional development sessions on using the frameworks to plan lessons.

In accounting for these findings, it is important to note that the effective use of pacing guides and curriculum frameworks both involve adjusting *what* is taught and *when* it is taught, but only the effective use of frameworks requires teachers to reorganize *how* they teach. As discussed in chapter 3, this latter type of development involves significant teacher learning and requires sustained support. In terms of the framework described in chapter 2 for assessing the potential of instructional improvement strategies, we would have predicted that most teachers would use pacing guides but not curriculum frameworks as intended with only limited support.

Given the limited prior research on teachers' use of curriculum frameworks, these findings provide additional insight into the potential usefulness of curriculum frameworks for instructional improvement. Together with Fred Newmann and colleagues, we conjecture district curriculum frameworks can be an important support for teachers' learning, but only if they are viewed as an element of a coherent instructional system.[8] This implies that the types of guidance the frameworks provide about, for example, instructional strategies, formative assessment, and differentiating instruction, should be a sustained focus of pull-out professional development, teacher collaborative meetings, and one-on-one coaching.

We would recommend that a district not develop curriculum frameworks unless it is willing to work to achieve this level of coordination. Their creation and ongoing revision requires a considerable investment of resources. For example, in one of our partner districts, two very capable district mathematics specialists worked on the curriculum frameworks fulltime. Our findings indicated that this investment of the specialists' expertise did not pay off.

Influence of Teachers' Use of Common Instructional Materials on Their Learning

To this point, we have emphasized that many teachers in our partner districts struggled to effectively implement the high-cognitive demand tasks included in high-quality instructional materials. However, it is also important to note that it is extremely difficult for teachers to increase the rigor of low-cognitive demand tasks. We found that across all districts and years, the level of rigor of only 3.7 percent of lessons was higher the potential of the tasks the teacher used. This finding indicates it is crucial for teachers to have access to high quality instructional materials if they are to have a chance at supporting students in attaining rigorous goals learning goals.

It was also apparent that the lack of common instructional materials that aim at rigorous learning goals both within and across schools impeded the development of coherent instructional systems at the school level. For example, we noted in chapter 5 that high-quality professional development is organized around the instructional materials that teachers are using. The lack of common materials that aim at rigorous learning goals makes it more difficult to provide high-quality, content-focused professional development. We found that across the MIST districts, most teachers reported in interviews that professional development sessions that focused on upcoming instructional units were the most useful. Often called "just-in-time professional development," these sessions gave teachers the opportunity to solve the tasks they would use in upcoming lessons, discuss solution strategies, and identify potential student misconceptions. District-level professional development of this type is not possible if teachers are using a range of different instructional materials. In the district that discontinued district-wide use of CMP2, professional development providers and instructional coaches spoke of the challenges of providing useful professional development because sessions could not be organized around upcoming tasks that all teachers would use.

We also found that using common instructional materials supported more productive teacher collaborative meetings. In the district that moved away from CMP2 and allowed schools to choose their own instructional materials, the amount of time that teachers spent searching for or developing their own materials took time away from collaborative activities that could support instructional improvement, such as collectively examining the tasks in upcoming lessons, discussing the underlying mathematical ideas, and identifying potential implementation challenges.[9] As discussed in chapter 4, coaches work with individual teachers in their classrooms is in turn likely to be more productive if it can build on teacher collaborative meetings in which teachers have planned lessons together.

TAKEAWAYS FOR INSTRUCTIONAL LEADERS

Here, we highlight key takeaways regarding instructional materials for instructional leaders (e.g., district leaders, school leaders, coaches) charged with designing and implementing strategies for supporting mathematics teachers' development of ambitious and equitable instructional practices.

Districts should adopt common, high-quality instructional materials. Taken together, our findings indicate the potential benefits of providing teachers with a common set of high-quality instructional materials across schools. Having access to high-quality instructional materials increases the likelihood that teachers will implement high-cognitive demand tasks. If it is not possible to adopt a single set of instructional materials, we recommend that instructional leaders with expertise in ambitious mathematics instruction and knowledge of students' mathematical learning review the instructional resources that teachers' often use and select high-quality mathematics lessons and units that support the district's vision of high-quality instruction and are sequenced to develop students' understanding of key mathematical concepts.

Teachers must be supported to use high-quality instructional materials effectively. We know from numerous studies that the adoption of high-quality curriculum materials does not automatically result in teachers' development of high-quality instructional practices.[10] Our findings about why some teachers substituted lower-level tasks for CMP2 tasks have direct implications for the design of pull-out professional development that is organized around instructional materials. As described in chapter 5, professional learning opportunities should address the challenges teachers face in maintaining the cognitive demand of rigorous tasks during instruction. Additionally, our finding that many teachers did not think that inquiry-oriented instructional materials were appropriate for some of their students further emphasizes the importance of supporting teachers' development of productive views of their students' current mathematical capabilities. Pull-out professional development, teacher collaborative meetings, and one-on-one coaching should include activities designed to address teachers' concerns about the appropriateness of ambitious instruction for all their students. These activities might also address how high-quality instructional materials align with the district vision for mathematics instruction and how they can be used to support all students in attaining rigorous learning goals.

FUTURE RESEARCH

We have identified two pressing areas for future research regarding instructional materials. First, there is still much to learn about how teachers interpret and use high-quality instructional materials.[11] For example, there is currently limited research on addressing teacher views of the appropriateness of inquiry-oriented instructional materials for all their students, and about supporting teachers to better assess what their students can currently do with appropriate support. A second area for future research focuses on clarifying the role of curriculum frameworks within a coherent instructional system that aims to support teachers to develop ambitious and equitable instructional practices. This issue is pressing given that many districts are investing heavily in the development of curriculum frameworks and it is not clear that those investments are paying off.

TEN

Educational Assessments

BRETTE GARNER, NICHOLAS KOCHMANSKI,
AND ERIN HENRICK

IN THIS CHAPTER, we focus on how teachers, school leaders, and district leaders can use assessments to support the development of ambitious and equitable instructional practices. Prior research has shown that assessments can help teachers clarify expectations for student learning and assess what students have learned.[1] We also know that well-designed assessments can inform pedagogical decision-making in ways that support more ambitious and equitable instruction.[2] For example, a middle-school mathematics teacher might identify an assessment task that is linked to state standards. By solving the task herself (and anticipating various student solutions to the task), the teacher has the opportunity to identify the knowledge and skills that are elicited by the task. She can then use this information to develop an instructional plan, including clarifying daily or unit-level goals for student learning. After students engage with the task, the teacher can examine their work to identify what students understand about the standards addressed by the task. The teacher can then plan future instruction that responds to and builds on student thinking, a hallmark of ambitious and equitable instruction. However, using assessments to inform instruction in these ways is not typical, particularly in the context of high stakes accountability policies.

The findings of numerous studies indicate that the pressure of high stakes accountability testing often distorts the use of assessments in ways that conflict with rather than support instructional improvement.[3] Tensions between the two uses of assessments—for instructional improvement and for accountability purposes—were evident throughout the MIST project. Although all four partner districts aimed to increase student achievement by improving the quality of math instruction, district leaders also felt a great urgency to improve test scores rapidly in order to meet the adequate yearly progress requirements in No Child Left Behind.

In this chapter, we begin by describing our partner districts' assessment initiatives. We then share findings related to two issues we investigated pertaining to district assessments: alignment with the district's rigorous goals for student learning, and the use of assessment data to support instructional improvement. In investigating these two issues, we drew on the expectations for how assessment data would be used as specified in the District Design Documents, and our findings about how assessments were actually used as reported in our annual District Feedback and Recommendations Reports (across all years and districts). We also reviewed released state assessment items and drew from video-recordings of teacher collaborative meetings in which teachers and coaches analyze data together (see chapter 6 for more detail). We conclude by sharing a set of takeaways for school and district leaders seeking to use assessments to support the development of ambitious and equitable instruction, and suggest areas for future research in this area.

PARTNER DISTRICTS' ASSESSMENT INITIATIVES

Assessments were a component of each district's improvement efforts in various years throughout the partnership. For example, one district aimed to use multiple forms of assessment for instructional planning and decision-making. Another district worked to align curriculum, assessment, and professional development. The other two districts used assessments to support the implementation of a rigorous curriculum. Three of the four districts developed benchmark assessments that were administered every three, six, or nine weeks depending on the district and year. The intent of these assessments was to gauge how schools and students were likely to perform on the end-of-year state assessment. Three of the four districts also launched district-wide formative assessment initiatives at different times during our work with them. As part of these various initiatives, the districts increased the number of school-level positions (e.g., data coach) devoted to preparing and/or analyzing student assessment data for teachers and administrators. However, although the districts invested substantial time and resources to develop and implement their assessment initiatives, they were not always aligned with other elements of their instructional systems.

ALIGNMENT BETWEEN ASSESSMENTS AND DISTRICT GOALS FOR STUDENT LEARNING

Our first set of findings concern the extent to which the assessments the districts used were aligned with their rigorous goals for instruction and student learning. We addressed this issue by examining state accountability assessments, district benchmark assessments, and classroom-level formative assessments.

Alignment Between State Assessments and District Goals for Students' Learning

As has been noted in prior chapters, for most of the years that we partnered with the districts, the state assessments emphasized procedural competencies at the expense of conceptual understanding and problem solving.[4] We found that most principals perceived that they were held accountable primarily for increasing student performance on these assessments, and only secondarily for supporting teachers' development of ambitious and equitable practices (see chapters 12 and 13). Principals in turn tended to communicate to teachers that they should focus on improving their students' performance on procedurally-oriented state assessments, thereby steering teachers away from the districts' rigorous goals for students' learning.[5]

As noted in chapter 2, the rigor of state assessments increased in the later years of the MIST project, thus improving their alignment with the districts' learning goals.[6] However, most school leaders and district leaders (with the exception of district mathematics leaders) did not realize that students would need to develop new mathematical capabilities if they were to perform well on these new, more rigorous assessments (see chapters 12 and 13). As all states have now implemented more rigorous standards and assessments, this latter finding points to the importance of supporting school and district leaders in understanding the implications of more rigorous assessments for students' learning and thus for instruction.

Alignment Between District Benchmark Assessments and Goals for Students' Learning

Three of our partner districts developed benchmark assessments that were intended to mimic state tests and thus typically included items of low cognitive demand that were poorly aligned with the districts' goals for student learning. In contrast, district math leaders in the fourth district used items from the *Connected Mathematics Project 2* (CMP2) item bank to construct benchmark assessments. The resulting assessments were therefore aligned with the district's rigorous student learning goals and could give teachers insight into the extent that students were developing conceptual understanding and problem-solving capabilities. However, the district's Leadership Department was concerned that the assessments would not adequately predict students' performance on the procedurally-oriented state assessment. Eventually, the district moved away from using CMP2 items and developed assessments that were less cognitively demanding and aligned with the state assessment.

Although interim benchmark assessments can help students become familiar with the types of items on state assessments, it is important for district leaders, school leaders, and teachers to be aware of their limitations. In all four districts, students who performed poorly on benchmark assessments were given additional support in the form of second mathematics classes or tutoring, which typically aimed to support students' development of procedural competencies (see chapter 11). Making decisions about the focus of the instruction that students should receive based on assessments that are not aligned with the rigorous learning goals that orient the overall instructional improvement effort is likely to impede students' attainment of those rigorous goals (see chapter 11).

Alignment Between Formative Assessment and Goals for Students' Learning

Three of our partner districts introduced formative assessment initiatives as part of their broader instructional improvement efforts. The intent of formative assessment is to use of evidence of students' reasoning (e.g., written responses to tasks, responses to in-class questions, contributions to discussions, etc.) to inform the ongoing adjustment of instruction and the planning of upcoming lessons.[7] Although teachers in all three districts received professional development on formative assessment, the majority of teachers in two of the districts described formative assessments as quick checks for understanding. For instance, teachers asked students to give a "thumbs up" or "thumbs down" to indicate whether or not they understood a topic. This practice provides teachers with information about whether students think they "got it" and thus whether parts of lessons need to retaught. However, quick checks of this type do not enable teachers to assess the depth of students' understanding of particular concepts or procedures and thus adjust their instruction accordingly.

In contrast, the third district implemented Formative Assessment Lessons (FALs) that were intended to be forward looking and to inform teachers' ongoing improvement of their instruction (see chapter 4).[8] These instructional units included cognitively demanding tasks for teachers to use as the basis for their lessons together with formative and summative assessments. These assessments were designed to elicit student understanding, thereby enabling teachers to adapt their instruction to students' learning needs.

The FALs were aligned with the district's ambitious agenda for mathematics teaching and learning. Further, teachers received high-quality pull-out professional development and coaching that was organized around the implementation of FALs (see chapter 4). Although there was considerable variation in how teachers used the FALs during the first years of implementation, this was the only instance in which we observed a significant number of teachers develop forward-looking formative assessment practices. The FALs example is useful as it indicates the types of aligned supports that most teachers are likely to need if they are to develop such practices, including rigorous instructional materials with embedded formative assessment tasks and sustained support for their professional learning that is organized around those materials.

DATA USE TO SUPPORT INSTRUCTIONAL IMPROVEMENT

Our second set of findings focus on how data can be used to support instructional improvement. As part of their assessment initiatives, leaders in all four partner districts expected teachers and coaches to analyze student assessment data. The purpose that district leaders gave for analyzing data was to inform instructional planning and decision making. However, what data analysis looked like in practice varied widely across schools and districts, and was usually unlikely to support teachers' development of ambitious and equitable instructional practices.[9] For example, our analysis of video-recorded teacher collaborative meetings indicates that most groups analyzed data from benchmark assessments, typically with the goal of identifying either content to re-teach or students who needed additional support. However, most teachers did not attempt to understand why students had performed poorly on particular items and thus how they might improve how they taught that content.[10] This is worrying given that we selected teacher groups that were more likely to work together effectively (see chapter 6 for our criteria for selecting teacher groups). We did, however, identify some workgroups that collected additional types of data and analyzed those data productively. We focus on these groups to illustrate how the use of data can support the development of ambitious and equitable instructional practices, and to clarify how instructional coaches and school and district leaders can support teachers' productive use of data.

Teachers' Data Use for Instructional Improvement

The productive cases typically involved an instructional coach asking teachers to bring students' work on assessment tasks that elicited student thinking to a workgroup meeting. As an illustration, the sixth-grade workgroup at Magnolia Middle School used district benchmark assessments to identify content that students found difficult and then developed common, open-response assessment tasks that elicited student thinking around that content. When the teachers analyzed students' written responses to the tasks, they worked to identify what students did and did not understand. In one meeting, for instance, they concluded that some students could accurately compute a unit rate, but that they did not yet grasp the deeper multiplicative relationship between the two quantities.[11] The teacher workgroup then designed future lessons that supported students' understanding of the underlying concept of unit rate.

Through their analysis of benchmark assessment data and student work, the Magnolia teachers were able to investigate students' current understandings and relate what they found to their prior teaching.[12] Several other teacher workgroups also conducted similarly productive analyses of students' responses to high cognitive demand assessment tasks like those found in the FALs. Although these practices were atypical across the MIST districts, these cases demonstrate how expanding the notion of data to include qualitative as well as quantitative data that give insight into student thinking can support instructional improvement.

Coaches' Data Use for Instructional Improvement

Prior research suggests that instructional coaches with expertise in ambitious and equitable instruction and in assessment can support teachers' data use for instructional improvement.[13] However, most school-based coaches in our partner districts spent a great deal of time preparing data to share with teachers and school leaders, and determining interventions for students based on assessment data (see chapter 7). These tasks greatly reduced the extent to which they could support teacher workgroups in analyzing data to improve. Consequently, teachers in these schools had access to large amounts of data but received little guidance on how to use the data productively.

We did, however, observe several cases in which a coach worked with teacher workgroups to use data thoughtfully.[14] At Magnolia, for example, the facilitators (an instructional coach and an assistant principal) developed a three-week cycle of assessment and data analysis that supported the teachers in analyzing data to improve instruction (see chapter 6 for a more detailed description of this cycle). The facilitators led the teachers in developing formative assessments tied to rigorous learning goals and in analyzing students' responses to the assessment tasks. The results of these analyses then informed the teachers' planning of future instruction. The facilitators' expertise was crucial for maintaining the group's focus on student thinking and on supporting students' development of conceptual understanding and problem-solving capabilities. At Magnolia and at the other schools where teachers' use of data was productive, facilitators helped manage the large amount of data available, but also maintained teachers' focus on using data to learn about their students' thinking and to use what they learned to improve their instruction.

School and District Leaders' Data Use for Instructional Improvement

School and district leaders can play an important role in supporting teachers' and coaches' data use. They influence the collection and flow of data by implementing assessment initiatives (e.g., benchmark assessments) and by selecting data management systems.[15] They also communicate expectations regarding the types of data that teachers and coaches should analyze and the purposes for which they should do so.[16]

Over the course of the MIST project, school and district leaders across all four districts consistently reported that using assessment data to inform decision-making was a key aspect of their instructional improvement efforts. However, many described using data in ways that were potentially at odds with the districts' ambitious agendas for teaching and student learning. For example, some school leaders responded to accountability pressures from their supervisors by focusing almost exclusively on increasing student achievement in the short term rather than improving the quality of instruction in the long term (see chapters 12 and 13). These school leaders typically encouraged teachers to re-teach content on which students had performed poorly. In contrast, other school leaders fostered uses of data that were aligned with their districts' improvement agendas by encouraging teachers to consider the reasons for students' difficulties and to adjust their instruction accordingly to support students' deeper conceptual understanding.

TAKEAWAYS FOR USING ASSESSMENTS TO IMPROVE INSTRUCTION

We offer the following takeaways to support district and school leaders' efforts to use assessment data to support instructional improvement. The lack of alignment between a district's assessment initiatives and its rigorous goals for students' learning are likely to impede rather than support teachers' development of ambitious and equitable instructional practices. Therefore, our first takeaway is that developing a coherent instructional system requires that **districts adopt or create assessments that are aligned with the district's goals for student learning**. We suggest that districts also attend to the alignment between benchmark assessments and instructional materials so that the resulting data can inform teachers' efforts to improve their implementation of the materials.

In cases where state accountability assessments focus primarily on procedural competence rather than rigorous students learning goals, district and school leaders should balance meeting accountability demands in the short term with supporting instructional improvement

in the long term (see chapter 13 for a detailed discussion of this issue.) In such cases, districts might adopt or develop interim assessments that assess both procedural competence and conceptual understanding, and that can therefore bridge between state assessments, instructional materials, and the district's goals for student learning.

Further, we suggest that **school and district leaders' primary role in fostering productive uses of data should be to organize supports for teachers to learn to use data to improve their instruction, including identifying accomplished facilitators for teacher workgroups.** It was not common for teachers in our partner districts to use assessment data to investigate and improve instruction. We therefore recommend sustained, job-embedded supports–including working with accomplished instructional coaches–on analyzing assessment data and using the resulting evidence about students' thinking to inform instructional planning and decision-making.

FUTURE RESEARCH

Although our findings provide some insight into how effective uses of assessment data can support instructional improvement, there are significant gaps in the research base in this area. First, we need to better understand which types of assessments and which contexts for analyzing the resulting data best support teachers in linking what they learn about their students' thinking to instructional decision-making. Although teacher collaborative time has the potential to support instructional improvement, research on how teacher workgroups can best make use of various types of data to improve instruction is clearly needed. Second, we do not know enough about what instructional coaches need to know and be able to do to support teacher workgroups to analyze data productively. Our finding that productive workgroups were the exception rather than the rule indicates that this is a pressing issue.

ELEVEN

Supplemental Supports for Currently Struggling Students

JONEE WILSON AND DENICE KELLEY

THIS CHAPTER[1] FOCUSES on the final element of a coherent instructional system, the provision of *supplemental supports for students who are currently struggling* in mathematics.[2] We limit our analysis to an extremely common type of supplemental support that schools in all four partner districts implemented, so-called "double dose" or second math classes. These classes were created for students who had been identified as not meeting grade-level standards on state assessments. Research indicates that in the United States providing a second class is the most common form of support for students identified as struggling.[3] For example, in a recent study on how districts are responding to the challenge of supporting all ninth grade students to be successful in algebra, June Mark and colleagues found that 79 percent of 235 district mathematics leaders from large districts in the US reported offering additional classes.[4]

We did not anticipate the prevalence of this strategy at the start of the MIST project. Yet, it quickly became apparent that we needed to attend to it given its relevance to a coherent instructional system organized around rigorous learning goals and an ambitious and equitable instructional vision.

This was a pressing issue because the goals and practices of the second math classes in our partner districts were not, for the most part, aligned with the learning goals and instructional vision that guided the districts' instructional improvement efforts. Moreover, consistent with US national data, many of the students identified as not meeting grade-level standards were members of groups that have been historically underserved.[5] The ways in which supplemental math classes were being implemented therefore impacted the opportunities for historically underserved groups of students to develop rigorous mathematical understandings.

Although the strategy of providing a second class for struggling students is common, the research base on which school and district leaders might draw regarding how to implement this strategy is remarkably thin, and the findings from albeit limited research is mixed.[6] Thus, it is not surprising that our partners struggled to implement supplemental classes in ways that adequately served their students. In what follows, we share a MIST analysis that focused on the goals of supplemental classes and on their effectiveness in terms of improving student achievement. We then highlight three key challenges that our partner districts encountered in implementing second math classes. In light of our findings, we share takeaways for district and school leaders about designing and implementing supplemental classes that are aligned with rigorous learning goals and an ambitious and equitable vision of instruction. Last, we suggest an agenda for future research.

LACK OF EFFECTIVENESS OF SUPPLEMENTAL CLASSES AIMED AT LOW-LEVEL LEARNING GOALS

In all four of our partner districts, decisions about how to support students identified as not meeting grade-level standards were made at the school level. All school leaders reported in interviews that their primary goal in implementing supplemental classes for students identified as "low-performing" was to increase the identified students' achievement scores on end-of-year state assessments. As described in chapter 2, for most years of the MIST project, the state assessments tended to target procedural capabilities with little attention to problem-solving and conceptual understanding. School leaders' goal of directly supporting students to perform better on procedurally oriented assessments differed slightly from the goals articulated by teachers. In interviews, most teachers of second math classes attributed students' low performance in large part to a lack of basic skills needed to complete the class activities successfully in their mainstream mathematics classes. Teachers thereby viewed the purpose of the supplemental classes as remediating these missing skills, and their descriptions of the classes indicated that they focused on isolated skills and procedures for the most part.[7] There was minimal evidence that supplemental instruction in an overwhelming majority of our partner schools would support the development of mathematical reasoning and problem-solving capabilities.

It is reasonable to expect that providing struggling students with additional instruction that focused on basic skills and procedures would lead to improvement in student achievement on procedurally oriented assessments. However, an analysis of the implementation of second math classes across all four districts in years one through four of the project revealed that the *supplemental classes were rarely effective in improving student achievement on state assessments that focused primarily at low-level goals.*[8] Specifically, an analysis that controlled for students' prior achievement and demographic characteristics indicated there was, on average, a negative or statistically insignificant difference in the achievement of students who were in the additional classes as compared to those who were not.[9] In fact, in only four of thirty schools did students with additional instructional time perform better than expected on state assessments, given their prior achievement. Moreover, we found that in schools that implemented supplemental classes, there were lower achievement gains than would have been predicted based on students' prior achievement for *all* students.[10]

KEY CHALLENGES IN PROVIDING SUPPLEMENTAL CLASSES FOR CURRENTLY STRUGGLING STUDENTS

The finding that supplemental classes rarely resulted in improved achievement on state assessments that were targeted at low-level goals is, at first glance, surprising. However, an analysis of interviews with school leaders and especially teachers helps explain the finding. We found that teachers did not have access to three supports that albeit limited research suggests matter: high-quality instructional materials organized around rigorous learning goals, purposeful collaboration across mainstream and supplemental classes, and professional development.[11]

Learning Goals and Instructional Materials

A key aspect of designing and implementing supplemental support classes concerns the identification of goals for students' learning and the selection of instructional materials (see Chapter 9). As highlighted above, the learning goals in most supplemental classes were at odds with the rigorous goals and associated vision for mathematics teaching that oriented mainstream classroom instruction and, more broadly, the districts' instructional improvement efforts. Attending to the quality of supplemental instruction was not initially in our scope of work, and we did not budget to video-record supplemental instruction. However, our interview data indicate that teachers rarely coordinated the focus of the supplemental math class with the focus of mainstream lessons. Consequently, students identified as needing supplemental support participated in two mathematics classes that aimed at decidedly different kinds of learning goals.

In addition, our interview data indicate that a majority of our partner schools and districts did not adopt a specific set of instructional materials for these classes. For example, in the first four years of our project, 85 percent of the schools in one of our partner districts did not adopt any materials.[12] This is concerning given the finding that, in the absence of an adopted set of instructional materials, additional instructional time is associated with lower than expected student achievement scores.[13] Many teachers reported that they primarily used computer software packages so that their students could work on different sets of exercises. Other teachers reported pulling worksheets from existing resources (e.g., the internet, available textbooks) and/or creating worksheets. In both cases, the activities focused on using procedures to solve routine types of problems.

Collaboration between Teachers of Mainstream and Supplemental Classes

In each of the four (out of 30) schools in years one through four in which additional time was associated with higher than expected student achievement for currently struggling students, the same teacher taught the mainstream and additional support class.[14] This finding highlights that additional instructional time is unlikely to advance students' learning unless deliberate connections are made between mainstream coursework and the supplemental instruction.[15]

However, less than half of the participating teachers in our partner schools taught both mainstream courses and supplemental courses, and only about one-third of these teachers taught the same students in both the mainstream and the supplemental class. This means that in over 80 percent of our partner schools, students received mathematics instruction from two different teachers. Furthermore, teachers reported that there was rarely any collaboration between teachers of the mainstream mathematics classes and of the additional support classes. In the absence of purposeful communication between teachers and aligned instructional materials, students learning opportunities in the two classes were likely disconnected.

Learning Demands for Teachers of Supplemental Classes and Professional Development

The available evidence indicates that teachers will likely need to develop new instructional practices if they are to support currently struggling students to develop rigorous understandings of mathematics. For example, research suggests that to support currently struggling students' development of robust mathematical understandings, it is imperative to identify the specific difficulties that students are having, and to provide instruction that takes account of and builds upon students' current understandings.[16] Doing so requires a deep knowledge of how students' understandings of key mathematical ideas develop, and the capacity to make sense of students' current understandings in light of this knowledge.[17] Moreover, research suggests that it can be useful to pre-teach the skills and understandings that will be required to participate effectively in the mainstream class. Pre-teaching involves "setting the stage" or providing an "orientation session."[18] The idea is to introduce central concepts that are crucial for tackling the challenging tasks that will be the focus of upcoming mainstream instruction. *Pre-teaching* should not be confused with *proceduralizing*. Proceduralizing involves teachers introducing specific procedures, thereby lowering the rigor of instructional tasks. In contrast, in a well-designed pre-teach lesson supports students' development of the capabilities necessary to participate fully in and learn from upcoming mainstream lessons. Thus, the intent of pre-teaching is to scaffold students' development of conceptual understanding without reducing the rigor of the tasks that students will tackle in mainstream instruction. Learning how to pre-teach without reducing the rigor of tasks is demanding work.

Our findings also suggest that supporting struggling students successfully requires that most teachers develop more productive views of their students' current mathematical capabilities. As reported in chapter 3, a sizeable number of teachers in all four MIST districts did not view students who they perceived as struggling as capable of engaging in rigorous mathematical activity. In light of these teachers' views of their struggling students' current capabilities, it is not surprising that supplemental classes focused primarily on "basic skills" and facility with procedures.

Developing the knowledge, perspectives, and forms of practice described above requires sustained professional learning opportunities. However, none of our partner schools and districts provided professional development for teachers of supplemental classes around designing lessons and developing instructional practices intended to support currently struggling students to meet rigorous learning goals.

KEY TAKEAWAYS FOR DISTRICT AND SCHOOL LEADERS

Supplemental classes can be a crucial support for currently struggling students. However, a number of implementation decisions influence whether these classes actually enable students to develop rigorous understandings of mathematics and problem-solving capabilities. In what follows we highlight key takeaways for district and school leaders who are providing second mathematics classes as a means of supporting currently struggling students.

Supplemental classes should be aligned with the rigorous learning goals and the corresponding vision of ambitious and equitable instruction that orient mainstream instruction. In other words, additional instructional time should be used to support students' development of the same robust and enduring conceptual understandings, and problem-solving capabilities, that are the focus of regular mathematics classes. Moreover, support classes should be designed to give students time to engage in problem-solving activities *that are directly related to the focus of the mainstream class*, rather than merely providing more time to practice mathematics skills that are not meaningfully connected to the mainstream lessons. And, any focus on procedures should include substantial attention to understanding how and why procedures work.[19]

This takeaway is informed by the findings reported above and reflects the stance that all students, especially historically marginalized groups of students, deserve opportunities to develop productive mathematical identities and robust mathematical understandings. Crucially, ensuring that supplemental classes focus on rigorous goals for students' learning that encompass problem-solving capabilities and mathematical reasoning appears essential for improving achievement on the more rigorous assessments that are being implemented in all US states. These assessments prioritize tasks that require students to analyze novel problems in order to figure out the procedures they need to perform.

Alignment of the instructional materials used in mainstream and additional support classes is vital. It is important to provide the teachers of supplemental classes with instructional materials that are designed to support students' development of robust mathematical understandings. The materials used in the supplemental classes should connect to and reflect the rigorous learning goals being addressed the mainstream class. This could mean using additional instructional time to extend the lessons taught in mainstream mathematics classes. And/or, the additional time might be used to pre-teach concepts, procedures, and skills that are required to engage substantially in upcoming activities in the mainstream lesson. In either case, materials for supplemental instruction should emphasize students' development of conceptual understanding, including the conceptual underpinnings of procedures.

Schedule the same teacher to teach the mainstream and supplemental classes, or schedule time for the mainstream and supplemental support teachers to collaborate regularly. If having the same teacher teach students for both classes is not possible, then it is important to provide regularly scheduled time for the teachers of the two classes to collaborate and co-plan. (Of course, the effectiveness of having the same teacher teach both classes is dependent on the quality of the teacher's instruction.) Co-planning and other exchanges between the teachers of the mainstream and the support classes are vital if substantial connections are to be made across the respective coursework. Also, by working and planning together, the teachers may identify specific sources of students' difficulties and can act on these insights. This work of coordinating instruction in the two classes could be done in the context of teacher collaborative meetings (see chapter 6).

Provide sustained professional learning opportunities that focus squarely on how teachers can productively frame and respond to the

challenge of supporting currently struggling students. We can imagine that district-wide pull-out PD might aim to clarify the goals of supplemental classes, and to establish a network of teachers across schools who are charged with supporting struggling students. This PD could support teachers in understanding that supplemental classes should reflect a set of rigorous learning goals and an ambitious and equitable vision of instruction. We also anticipate it will be important to press and support teachers to identify *instructional reasons* for why students are struggling (rather than reasons that locate the source outside of teaching, such as inherent characteristics of the students, or their family or community backgrounds). This is essential in supporting teachers to develop more productive views of their currently struggling students' mathematical capabilities.

It is important that other types of support are aligned with the proposed pull-out professional development (see chapter 4). For example, teachers might work in teacher collaborative meetings to identify their struggling students' difficulties and develop plans for addressing these difficulties (see chapter 6). Teacher collaborative meetings also provide opportunities for teachers of supplemental and mainstream classes to identify the concepts and skills that should be pre-taught in supplemental classes so that currently struggling students can participate fully and learn from upcoming mainstream lessons. In addition, coaches could work one-on-one with teachers of supplemental classes to support their implement lessons that reflect an understanding of why students are struggling with particular concepts, and their pre-teaching concepts central to upcoming mainstream lessons.

FUTURE RESEARCH

As a research community, we know very little about how supplemental mathematics classes that aim to support struggling students' attainment of rigorous learning goals should be designed and implemented. As a consequence, research can currently provide school and district instructional leaders with only limited guidance as they make decisions about, for example, instructional materials for supplemental classes, teacher assignments, and the types of professional learning opportunities to offer teachers. It is imperative that researchers investigate these and related issues if the widely implemented strategy of providing second math classes is to be effective is enabling currently struggling students to attain rigorous learning goals.

One key area for future research concerns studies of the forms of expertise that teachers of supplemental classes need to develop if they are to support struggling students effectively. For example, how do successful teachers go about identifying their struggling students' specific difficulties? How do successful teachers determine what to pre-teach in relation to upcoming mainstream lessons? What specific instructional practices do they enact when they pre-teach particular concepts and skills, thereby enabling struggling students to participate fully in mainstream instruction?

Research is also needed on the design and implementation of professional learning opportunities that can support teachers of supplemental math classes in developing the targeted forms of expertise. For example, what should be the focus of district-wide pull-out professional development, teacher collaborative time, and one-on-one coaching, and how should these different types of support be coordinated? In the absence of this kind of research base, it is likely that schools will continue to implement supplemental classes that result in, at best, minimal beneficial outcomes, and, at worst, detrimental outcomes for currently struggling students.

TWELVE

School Instructional Leadership

CHARLOTTE SHARPE, ADRIAN LARBI-CHERIF,
ERIN HENRICK, PAUL COBB, AND THOMAS M. SMITH

AS DESCRIBED IN THE PRECEDING CHAPTERS, a coherent instructional system is oriented by rigorous learning goals for all students and an associated vision of high-quality instruction, and includes a coordinated system of supports for teachers' learning.[1] Although district leaders play a central role in designing this system, school leaders are primarily responsible for initiating and guiding the development of coherent instructional systems at the school level. Prior research indicates the importance of school leaders helping to establish a shared vision for improvement at the school level.[2] Additionally, school leaders can organize the school day, broker professional development, and allocate resources to support teacher learning.[3] Although findings from prior research help define the role of instructional leaders, the research does not provide clear guidance as to what school leaders actually need to know and be able to do to design and implement coherent instructional systems that support teachers' development of ambitious and equitable instructional practices.

We addressed this issue by investigating relationships between the types of improvement strategies school leaders implemented in their buildings, their visions of high-quality mathematics instruction, and the types of support that they received to improve their leadership practices. In doing so, we took account of the tasks that district leaders expected school leaders to carry out, and assessed whether it was feasible for districts to support school leaders in accomplishing these tasks effectively.

In this chapter, we first distinguish between school leaders working with teachers *directly* to support their learning, and school leaders supporting teachers learning *indirectly* by creating conditions for others with instructional expertise to work directly with them to support their learning. Our findings indicate that even with considerable professional development, it is extremely challenging for principals to be effective in directly supporting teachers' development of ambitious and equitable instructional practices. This finding is significant as it indicates that district initiatives that primarily rely on school leaders directly supporting instructional improvement are unlikely to be successful (e.g., observing classroom instruction and providing feedback to teachers). It appears more feasible for school leaders to support instructional improvement indirectly by creating sustained opportunities for teachers to work with others with expertise in ambitious and equitable mathematics instruction, such as district math specialists and accomplished coaches. As we discuss below and in chapter 13, these findings have implications for district leaders charged with supporting and monitoring school leaders.

EXPECTATIONS AND FINDINGS FOR SCHOOL LEADERS' PRACTICES

The central office leaders of our partner districts expected school leaders to improve instruction by carrying out two kinds of leadership tasks: directly supporting improvement by observing teachers and giving them feedback on their instruction, and indirectly supporting improvement by scheduling and communicating expectations for teacher collaborative meetings, and by working with coaches.

Supporting Instructional Improvement Directly: Observation and Feedback

At the beginning of our work with the four districts, our Theory of Action included the possibility of a *direct* role for school leaders in supporting instructional improvement. However, we also questioned whether school leaders without a mathematics background could develop a sufficient understanding of ambitious and equitable instruction to enable them to support their teachers' learning directly. Our partner districts' strategies for instructional leadership gave us the opportunity to examine this issue. School leaders in all four districts were expected to spend two hours or more in classrooms each day and to provide teachers with feedback about the areas that they should aim to improve. Observation and feedback was, and continues to be, a widely advocated leadership practice.[4] For this reason, we considered it important to investigate both its potential for supporting instructional improvement, and to identify the forms of support that school principals would need in order to enact this practice effectively. In doing so, we reasoned that for observation and feedback to be effective, school leaders would need to develop both a relatively sophisticated vision of ambitious and equitable instruction against which they could compare the instruction they observed and effective feedback practices.

Three of the four MIST districts supported school leaders' learning by joining with a nationally known school improvement organization to provide professional development focused on broad principles of high-quality instruction that are applicable across content areas and on what to look for in classrooms. These principles included, for example, that teachers should communicate clear expectations to students and use rigorous instructional tasks as the basis for their instruction. Professional development sessions about this latter topic included examples and discussion of what rigor would look like in various content areas. School leaders in these three districts were also expected to observe instruction in multiple classrooms regularly, joined by teachers, coaches, and district leaders. These group observations were intended to help school leadership teams assess the degree to which teachers were integrating what they were learning from teacher professional development into their instruction. Mathematics was not the sole focus of these walkthroughs, though principals were expected to observe mathematics classrooms and look for the indicators of high-quality instruction. Given the districts' substantial investments in school leader professional development, we investigated the extent to which school leaders in these districts were giving teachers feedback that had the potential to support improvements in their instruction.

In this investigation, we focused on the relationships between school leaders' visions of high-quality mathematics instruction (as detailed in chapter 3), what school leaders attended to when observing instruction, and what feedback they subsequently gave teachers. When we analyzed the twenty-nine participating school leaders' visions of high-quality mathematics instruction, we found that many described several *forms* of inquiry-oriented math instruction (e.g., real world tasks, teacher questioning, use of manipulatives). However, few school leaders could explain the intended *function* of these aspects of instruction in supporting students' learning and, as a consequence, did not distinguish between weak and strong inquiry-oriented instruction. We also found school leaders' form-level visions of high-quality instruction were relatively stable over time, though we did see some improvements when school leaders received substantial math-specific professional development.[5]

The stability of school leaders' instructional visions, even when three of the four districts invested heavily in content-general instructional leadership professional development, raises questions about whether this type of professional development can enable them to provide effective feedback after observing mathematics classrooms. Although the vision of high-quality instruction undergirding this professional development was broadly compatible with the vision of high-quality mathematics instruction outlined in chapter 3, most school leaders were not able to translate it into a mathematics specific vision of instruction for their school, at least without additional support from those with expertise in mathematics instruction.

Having assessed school leaders' instructional visions, we sought to understand what feedback they were giving teachers after observing their classrooms. We therefore asked teachers in annual interviews and online-surveys how frequently they were observed by their administrator(s) and what feedback they received afterward (if they received feedback at all).

We found that the majority of the feedback teachers received was unlikely to support their development of ambitious and equitable instructional practices. Most teachers reported that feedback focused on readily observable aspects of instruction (e.g., whether an objective was posted on the board) or content-general aspects of high-quality instruction (e.g., students should work together in groups). Fewer than a quarter of the teachers mentioned getting feedback from their school leader that was specific to mathematics instruction; of those who did, only a few described receiving feedback that specifically addressed areas for improvement based on the specific lesson the leader had observed.[6] Moreover, we found no relationship between how often a teacher was observed and received feedback, and whether their instruction improved from one year to the next.

These findings were disappointing given the emphasis that all four districts placed on observation and feedback. The findings call into question the effectiveness of content-general professional development for helping school leaders learn what to look for in classrooms to determine whether mathematics instruction is of high quality. Moreover, the findings indicate that expecting school leaders to directly support teachers by observing and providing feedback on their instruction is unlikely to improve teachers' development of ambitious and equitable instructional practices.

Supporting Instructional Improvement Indirectly: Coordinating Supports for Teachers' Learning

School leaders in our partner districts were also expected to support instructional improvement *indirectly* by coordinating building-level supports for teachers' learning, in the process guiding the development of a productive teacher learning subsystem. Examples of indirect instructional leadership included:

- Scheduling and protecting time during the school day for teachers to collaborate on instructional issues.
- Asking instructional experts at the district level or within the school to support teacher learning by leading teacher professional development, facilitating teacher collaborative meetings, or providing one-on-one coaching.
- Identifying teachers who had already developed high-quality instructional practices and giving them leadership responsibilities (e.g., mentoring novice teachers).

In contrast to observation and feedback, school leaders do not directly attempt to support teachers in improving their instruction in the above examples. Instead, school leaders act as organizers who create conditions for teachers' learning, ensuring that teachers have opportunities to work directly with people with expertise in ambitious and equitable instruction.

Creating Opportunities for Teacher Collaboration

School leaders in all four partner districts were expected to create time in the school day for teacher collaborative meetings. Our interviews with teachers and school leaders indicate that most did so. However, as reported in chapters 6 and 10, school leaders frequently expected teachers to use this time to coordinate the pacing of lessons or use student assessment data to identify students for supplemental supports such as second mathematics classes or tutoring. While pacing is an important aspect of planning, making it a major focus of collaborative meetings without also considering the quality of instruction and student learning is unlikely to support teachers in developing ambitious and equitable practices.

In contrast to the kinds of expectations that most school leaders communicated for teacher collaborative meetings, a synthesis of the relevant literature suggests that it is important that school leaders ensure that collaborative meetings focus on key instructional issues.[7] This includes making it clear that assessment data can and should be analyzed to inform teachers' ongoing improvement of their instruction, not only to identify students who are currently struggling. Consistent with findings reported in chapters 6 and 7, it is also clear that school leaders should ensure that teacher collaborative meetings are facilitated by someone with instructional expertise and with expertise in supporting teachers' learning, such as an accomplished mathematics coach.

Collaborating with Coaches

As described in chapter 7, the four districts' coaching designs varied. In some districts, district-based coaches served several schools and were accountable to district mathematics specialists. In other cases, mathematics coaches were school-based and reported to the principal. The district- and school-based coaching models entailed tradeoffs. Although district-based coaches had greater autonomy and were usually not assigned additional tasks that reduced the time they could work with teachers on instructional issues, they were also less embedded in the schools they worked with and often found it challenging to build trust with teachers (see chapter 7).[8]

Prior research on mathematics coaching indicates that teachers are more likely to work with coaches and be open to learning from them when their principal works to position the coach as a support for teachers' learning, rather than as an evaluator of their instruction.[9] We built on this research by investigating how school leaders and coaches in the four districts worked together and toward what goals. We asked coaches in annual interviews what their school leaders expected them to be doing, how they were held accountable for this work, and in what ways they interacted with their administrators.

Our analysis indicates that there was wide variation across districts in school leaders' expectations for mathematics coaches. Some school leaders expected coaches to perform administrative duties and attend—but not facilitate—teacher collaborative meetings. Others expected coaches to observe instruction, work one-on-one with teachers, and facilitate teacher collaborative meetings.

In a few cases, school leaders and coaches worked together to support teachers' learning (see chapter 7). Although we did not see this type of collaboration between coaches and principals very often, an in-depth analysis of these cases suggests that characteristics of productive collaborations include:

- School leaders communicate that the intent of teacher collaborative meetings is to support improvements in the quality of instruction, and that the coach should engage teachers in activities that have the potential to support their development of ambitious and equitable instructional practices;
- School leaders and coaches observe classroom instruction together to assess teachers' practices, identify the focus of future professional development and teacher collaborative meetings, and determine which teachers need one-on-one classroom support from the coach; and
- School leaders and coaches meet regularly to discuss what they have observed in teachers' classrooms and determine how to support particular teachers.

These findings resonate with prior research indicating that principals who actively support the work of the coach can influence the coach's legitimacy among teachers.[10]

Fostering Productive Teacher Advice Networks

Thus far, we have discussed the crucial role of school leaders in designing and implementing the various facets of a teacher learning subsystem, including teacher collaborative time and coaching. As described in chapter

8, teachers' advice networks are also an important support for their learning. Importantly, school leaders can influence teachers' networks through their expectations for instruction, through positioning the coach as a source of instructional expertise, through scheduling and structuring teacher collaborative meetings, and through teacher hiring and placement. One of our analyses of teacher networks in the participating schools highlighted that coaches' and principals' joint work can be consequential for how many teachers sought advice from coaches about math instruction, and how often they did so.[11] Another analysis indicated that school leaders' practices can impact the kind of expertise that is valued, and thereby influence who teachers turn to for advice.[12] For example, in one school a principal perceived that he was increasingly held accountable for improving student achievement on procedurally oriented state assessments. He, in turn, communicated that teacher collaborative meetings should focus on improving student performance on district and state assessments rather than improving the overall quality of instruction.[13] Analyses of audio recordings of teachers' conversations in teacher collaborative meetings revealed that across the three years, conversations shifted away from instructional issues and towards increasing students' achievement. Moreover, an analysis of teachers' advice networks in this school revealed that over the course of three years, teachers increasingly sought advice from the colleague with the highest student achievement scores rather than the school coach, who was the most expert in terms of her mathematical knowledge for teaching and the quality of her classroom instruction.

Accounting for School Leaders' Expectations for Teacher Collaborative Time and for Coaches' Work with Teachers

Our partner districts' focus on rigorous goals for students' learning and an ambitious and equitable vision of instruction implied that school leaders should organize their schools to support substantial, long-term improvements in the quality of instruction. At the same time, school leaders were responsible for improving student achievement in the short-term. And, as described in chapter 10, for most years of the MIST project, state assessments mainly evaluated students' proficiency in applying procedures to solve routine problems, with little attention to conceptual understanding or problem-solving capabilities.

As the illustrations we have given indicate, there was some variation in the extent to which school leaders attended to both long-term instructional improvement and short-term improvement in student achievement. An analysis of seventy-one principals' practices indicated that about two-thirds of principals implemented only strategies aimed at quickly boosting student performance on procedurally oriented assessments, meaning that they gave little attention to long-term instructional improvement.[14] For example, we described above the tendency for school leaders to suggest that teacher collaborative meetings should be used primarily to analyze student assessment data in order to identify students for tutoring (without also considering why, instructionally, students might not have performed well in the first place). We have also described the tendency for school leaders to direct coaches away from supporting teachers to improve the quality of their instruction in favor of organizing supplemental supports for students who had performed poorly on state and district benchmark assessments. And, as we report in chapter 11, school leaders often implemented second math classes for these students that focused primarily on procedures at the expense of mathematical reasoning and problem solving.

An analysis of school leader interviews revealed that the variation in the strategies that school leaders implemented appears to depend on whether they viewed advancing student learning as requiring improvements in the overall quality of instruction.[15] School leaders who held this view tended to communicate that teacher collaborative meetings should focus on instructional issues and tended to ensure that coaches' work with teachers focused on supporting improvements in instruction. Importantly, our analysis revealed that school leaders who worked to improve the quality of instruction were more likely to recognize that instructional expertise was a critical support for improvement in their buildings.[16]

SUPPORTING SCHOOL LEADERS' LEARNING

Given the critical role that school leaders play in supporting the development of coherent instructional systems at the school level, it is imperative to consider the forms of support and guidance school leaders need if they are to foster conditions that support teacher learning, including professional development and relations with central office leaders.

Professional Development for School Leaders

An analysis of the interviews we conducted with school leaders found evidence that pull-out professional development can influence school leaders' effectiveness in supporting instructional improvement indirectly.[17] Specifically, the analysis revealed that the minority of principals who implemented strategies aimed at long-term instructional improvement had participated in professional development specific to inquiry-oriented mathematics and/or worked closely with an accomplished coach or a district mathematics specialist. These supports helped them in coming to value ambitious and equitable instruction and in viewing instructional improvement as requiring substantial teacher learning.[18] Consequently, these principals recognized and sought out the expertise of those who knew *how* to support teachers' learning. Importantly, we also found that recognizing and capitalizing on the expertise of colleagues who are capable of supporting teachers' learning does not require a sophisticated vision of high-quality math instruction.[19] This is good news as it lowers the threshold for what school leaders need to know about ambitious and equitable instruction, and about teachers' learning and how to support it.

Taken together, these findings suggest that school leaders should be provided with professional development aimed at supporting them to approach advancing student learning in terms of improving the quality of instruction, and in understanding that these instructional improvements involve substantial teacher learning. These two foci are especially relevant in light of the adoption of more rigorous standards and assessments in the majority of US states, as compared to the recent past.

The findings of another set of analyses are also relevant as they indicate the value of principals attending mathematics-specific professional development with the coaches who serve their schools.[20] This professional development aimed to support school leaders in distinguishing between high- and low-level math tasks, and we found that it positively influenced some of the school leaders' visions of high-quality instruction.[21] Importantly, a follow-up analysis indicated that this professional development also resulted in some of the school leaders working with instructional experts to design and implement supports for teachers' learning.[22] For example, principals who attended this professional development directed their coaches to lead pull-out teacher professional development that focused on key instructional practices. We conjecture that an aspect of the professional development that may have been important was that principals participated with the coaches who served their schools. This might have helped principals understand that developing ambitious and equitable instructional practices is challenging and requires sustained support and that their coaches are sources of expertise who could work directly with mathematics teachers.

Relationships with District Math Specialists

In addition to professional development, our analyses suggest that school leaders need ongoing support from mathematics content specialists in their districts. Even school leaders who recognize the need for teacher collaborative meetings to be facilitated by individuals with instructional expertise often require the support of a district mathematics specialist to help them identify teachers in the building with that expertise. Our analyses suggest that several principals who implemented strategies that had the potential to support teacher learning often sought out the district mathematics specialists for advice on how to improve instruction.[23] This typically included inviting district specialists to the school to observe mathematics instruction and provide recommendations for the types of support that teachers needed. In some instances, the district mathematics specialists also led pull-out professional development sessions at the school to support teacher learning.

Relationships with Principal Supervisors

We found that the expectations of the district leaders who worked directly with principals in their schools tended to influence how principals approached advancing student learning. A recurring issue in the feedback that we gave to our four partner districts each year concerned the lack of alignment between what school leaders perceived themselves to be accountable for and the districts' goals for instructional improvement. For example, if principals perceived that they were primarily held accountable for making significant gains on state assessments, then they were likely to primarily implement strategies such as procedurally oriented "double dose" or second mathematics classes and tutoring rather than strategies aimed at improving the quality of mainstream instruction. As we elaborate in chapter 13, these findings suggest that the expectations that principal supervisors communicate to school leaders should emphasize improving the quality of instruction in the long term as well as increasing student achievement in the short term.

TAKEAWAYS FOR DISTRICT LEADERS

School leaders should collaborate regularly with accomplished coaches and district math specialists to establish goals and design supports for improving instruction. Given that teachers require sustained support if they are to develop ambitious and equitable instructional practices, it is essential that principals collaborate with content experts to design and implement supports for teachers' learning. These ongoing collaborations might support principals' learning while simultaneously legitimizing and supporting the work of those with relevant expertise. For example, principals might develop more sophisticated visions of high-quality instruction and identify or at least vet professional development offerings for teachers based on their alignment with district instructional improvement goals. Furthermore, ongoing collaborations between math specialists and school leaders might result in teacher collaborative meetings and coaching being more productive, which could in turn positively influence teachers' advice networks (see chapters 6, 7, and 8). Principals, in collaboration with mathematics specialists or coaches, would then have taken a significant step in guiding the development of coherent instructional systems in their schools.

Reframe the purposes of administrator observation and feedback. Our findings suggest that it is likely not feasible for districts to support school leaders in giving teachers instructional feedback that is likely to support teachers' development of ambitious and equitable practices. We nonetheless still consider it important for school leaders to observe instruction regularly in order to have a presence in and learn what is going on in classrooms, as this should enable them to work more effectively with mathematics specialists or coaches. Furthermore, principals could use their positional authority to communicate instructional expectations that are consistent with district goals for instructional improvement, thereby legitimizing mathematics specialists' and coaches' work in directly supporting teachers' learning.

Support School Leaders in understanding connections between student learning, teaching, and teacher learning. Rather than expecting school leaders to directly support teacher learning through observation and feedback, district leaders should instead focus on supporting school leaders in understanding:

- that students need to develop conceptual understanding and problem solving capabilities if they are to attain more rigorous learning goals and perform well on more rigorous assessments,
- that these student learning goals have implications for what counts as high-quality instruction,
- that teachers' development of these instructional practices involves significant learning, and
- that teachers therefore require sustained support from others with expertise in ambitious and equitable instruction and in supporting teachers' learning.

As an illustration of how this might be accomplished, we collaborated with the mathematics specialists in one of our partner districts to design and lead a series of professional development sessions for school leaders that began by analyzing the demands of the more rigorous state assessments that had been recently introduced.[24] School leaders compared prior procedurally oriented assessment items with more rigorous items that assessed reasoning, problem solving, and conceptual understanding, and quickly realized that students would need to develop new capabilities if they were to perform well on the new assessments. We then traced the implications of the new standards and assessments first for teachers' instructional practices and then for the types of support that teachers would need to develop these instructional practices.

FUTURE RESEARCH

Drawing on the findings we have shared, we suggest three directions for future research on school leaders' role in supporting instructional improvement in their schools. First, additional research is needed to identify specific indirect leadership practices that can support teachers' development of ambitious and equitable instructional practices. The findings of such studies would further clarify goals for school leaders' learning, thereby orienting the design of professional development and other types of support, including ongoing collaboration with mathematics specialists or coaches.

Second and relatedly, there is a need for further research into how district mathematics specialists can collaborate productively with school leaders. Our findings indicate that these relationships can be an essential support to school leaders while also enabling mathematics specialists to be more effective.[25] However, little current research can inform district leaders about *how* they can foster productive collaborations between school leaders and subject matter specialists.

Finally, there is a need for additional research that investigates how school leaders can be supported to balance the short-term goal of improving achievement on state assessments with the long-term goal of improving the quality of teachers' instructional practices. School leaders will need support to see how supporting teachers' development of ambitious and equitable instruction and increasing student test scores are not incompatible goals. Without such support, the pressure to raise test scores in the short term might overpower districts' efforts to support all students' attainment of rigorous learning goals by improving the quality of instruction in the long term.

THIRTEEN

District Instructional Leadership

KARA JACKSON, PAUL COBB, JESSICA G. RIGBY, AND THOMAS M. SMITH

IN THIS CHAPTER[1], we turn our attention to the final component of our theory of action for instructional improvement at scale: district leadership. Over the past forty years, there have been significant changes in the relationship between district central offices and instructional improvement. For many years, central offices were considered bureaucratic, inefficient, and politicized organizations that had little to do with what happened in schools other than ensuring compliance with state and federal regulations.[2] More recently, however, central offices have shifted their focus from managing schools to supporting instructional improvement across the district.[3] The central offices of our partner districts reflected this trend; indeed, interviews conducted with central office leaders indicate that most were pressing for and attempting to support instructional improvement across schools.

Our leading contention is that *a primary goal of district instructional leadership should be to support the development of coherent instructional systems at the school level*, thereby building school-level capacity for instructional improvement. In this chapter, we share key findings regarding what is necessary for district leaders to achieve this goal, and discuss the challenges that our partner districts encountered. These findings, as well as relevant research literature, anchor the takeaways for central office leaders. In general, we found that the existing research was thin regarding what central office leaders need to *know and do* in order to support the development of coherent instructional systems, especially systems organized around rigorous goals for students' learning and a vision of ambitious and equitable instruction.[4] We conclude by discussing areas for future research on district instructional leadership.

We conducted interviews with district leaders in positions and offices that have the most direct involvement in mathematics teaching and learning. This included the superintendent; the chief academic officer (CAO), who is usually responsible for overseeing all matters relating to teaching and learning; as well as offices responsible for leadership (i.e., the office responsible for supervising principals), curriculum and instruction (C&I; i.e., the office responsible for supporting teachers, providing curricular materials, etc.), English learners, special education, and research and evaluation. In this chapter, we narrow our focus to senior district leaders (e.g., superintendent, CAO) and to leaders associated with the offices of leadership and C&I, given our findings regarding the importance of their role in all four partner districts' instructional improvement efforts. We recognize that there are differences between districts in terms of the influence that various departments have on the design and implementation of instructional improvement strategies. However, in what follows, we share findings that appear to cut across the four districts. In applying our findings, it might therefore be necessary to consider additional departments depending on the district context. For example, in one of our partner districts, the office of special education was especially influential and communicated a vision of instruction that was at odds with the vision of ambitious and equitable instruction articulated by leaders in C&I. In this case, it would be essential to take account of the role of leaders in this department as well as of those in C&I and leadership.

SUPPORTING THE DEVELOPMENT OF SCHOOL-LEVEL CAPACITY FOR INSTRUCTIONAL IMPROVEMENT

The Impact of Conflicting Agendas Between C&I and Leadership

One major challenge in supporting the development of coherent instructional systems at the school level concerns our finding that the offices of leadership and C&I frequently tend to pursue conflicting agendas aimed at advancing middle-grades mathematics learning.[5] Moreover, as we elaborate below, we found that when leadership and C&I pursued conflicting agendas for improving mathematics learning, there were consequences for all aspects of the instructional system.

The offices of leadership in our partner districts comprised a head of leadership (e.g., deputy superintendent, associate superintendent of secondary schools) and principal supervisors. Supervision included both formally evaluating the principals and providing them with ongoing support. Across our partner districts, principal supervisors were expected to hold principals accountable for supporting improvements in classroom teachers' instruction by, for example, observing and providing feedback to teachers and scheduling time for teachers to collaborate on a regular basis. They were also expected to hold principals accountable for increasing student achievement on state assessments.

The districts' C&I offices included a mathematics department that was charged with supporting teachers to improve the quality of their instruction by, for example, writing and updating curriculum frameworks and providing district-wide and school-based professional development. In addition, district-based mathematics coaches were members of the mathematics department and were responsible for providing support to teachers. Members of the mathematics department also provided some professional development for principals and assistant principals that was specific to a vision of ambitious and equitable mathematics teaching, although the amount of time allocated each year was typically small.

In the interviews that we conducted with them, the heads of the office of leadership and the majority of principal supervisors in all four districts described holding principals accountable for improving the quality of instruction. For example, they discussed communicating to principals the importance of ensuring that the high-quality instructional materials and curriculum frameworks or pacing guides provided by C&I were being implemented, and of observing classroom instruction regularly and providing feedback to teachers regarding the quality of instruction. While all of the principal supervisors discussed the importance of improving student achievement, they did not, at least in interviews, suggest that it was at odds with improving the quality of instruction.

In contrast, principals in all four districts reported in interviews across multiple years of the study that their supervisors primarily held them accountable for improving student achievement on district and state assessments, and only secondarily for improving the quality of instruction.[6] In other words, principals' reports indicate that principal supervisors communicated expectations that prioritized a short-term goal of boosting student achievement at the expense of a long-term goal of building school capacity for instructional improvement. It appeared, then, that while principal supervisors were in the position to press principals to improve the quality of instruction, they often did not.

This was especially problematic in the first four or so years of our partnership work because the state assessments at the time privileged relatively low-level goals for students' learning. Most principals therefore perceived that they should focus on increasing student achievement on assessments that required students to demonstrate proficiency in using procedures to solve routine types of mathematics problems. Meanwhile, C&I leaders provided professional development and instructional materials for teachers that were designed to support the development of students' conceptual understanding of central mathematical ideas and problem-solving capabilities.

As discussed in chapter 10, improving student achievement on state assessments and improving the quality of instruction are not, in principle, incompatible. Ample research has shown that students who receive ambitious instruction organized around inquiry-oriented instructional materials, like *Connected Mathematics Project 2*, tend to outperform students who receive procedurally oriented instruction on assessments of conceptual understanding and problem solving, and to perform as well on procedurally oriented assessments.[7] In addition, as the rigor of state mathematics standards and assessments has increased in all states, it is important for district leaders to appreciate that ambitious and equitable instruction is essential if student achievement on the new state assessments is to improve.

In what follows, we elaborate on the impact of the conflicting agendas between leadership and C&I on the development of coherent instructional systems at the school level.

This conflict is especially clear in the case of teacher collaborative meetings. An analysis of year two interview data in one district revealed that in schools in which school leaders perceived that their supervisors primarily held them accountable for increasing student achievement, teachers reported that the majority of the time in collaborative meetings was spent on test preparation activities (e.g., creating test-formatted warm-ups, planning how to teach particular test items).[8]

The influence of the conflicting agendas between leadership and C&I on a second element of a coherent instructional system, instructional coaching, was also relatively clear. As described in chapter 7, C&I intended that school-based coaches would provide a much-needed source of instructional expertise for teachers, the majority of whom were in the early stages of developing ambitious and equitable teaching practices. However, an analysis of school-based coaching in one district revealed that principals expected coaches to focus on improving student test scores. As they reported to principals rather than to C&I, they were frequently assigned tasks aimed at improving student performance on district benchmark and state assessments (e.g., analyzing student data, organizing tutoring).[9] This resulted in coaches spending less than 50 percent of their time working directly with teachers on improving the quality of their instruction. Coaches were given these additional responsibilities even though principals reported in interviews that they understood that coaches' primary responsibility should be to work directly with teachers on instructional issues.

Thus far, we have focused on how conflicting agendas between leadership and C&I impact the focus and quality of supports teachers receive at the school level (e.g., teacher collaborative time and coaching). We also found that conflicting agendas, mediated by school leaders' perceptions of what they were being held accountable for by their principal supervisors, can influence how instructional time is used. For example, in one of our partner districts, teachers who worked in schools where principals perceived that they were held primarily accountable for increasing student achievement reported that their principals expected them to spend half of each class period preparing for the state assessment and half using inquiry-oriented instructional materials.[10]

We also found evidence that conflicting agendas between leadership and C&I could impact the quality of supplemental supports for students identified as struggling. This was most evident in one of our partner districts in which a principal supervisor espoused that the rigorous learning goals associated with the district's vision were not appropriate for African American students who performed poorly on the state's assessment. It was no surprise, then, that the district text, *Connected Mathematics Project 2*, was used only minimally in the schools she supervised that served primarily African American students. Instead, school leaders directed their teachers to use instructional materials that emphasized procedural competence absent conceptual understanding. Moreover, this principal supervisor championed a policy implemented in all district middle schools that provided principals with funds to hire a teacher to provide additional mathematics classes for students identified as low performing on the state assessment. However, the district did not select instructional materials for the classes and did not provide professional development for the teachers.[11] Teachers' descriptions of the additional support classes in interviews revealed that instruction was primarily aimed at low-level student learning goals.

Accounting for Conflicting Agendas: Instructional Improvement and Instructional Management Orientations

As illustrated above, the conflicting agendas that we documented between leadership and C&I impacted multiple aspects of instructional systems at the school level. Here, we share our findings regarding what might account for these conflicting agendas. Based on an analysis of the annual interviews conducted with district leaders in all four districts, we found that the differences in agendas could be attributed, at least in part, to differences in how central office leaders—especially in leadership and C&I—framed the problem of advancing student learning.[12] Framing entails both identifying the source of the problem and suggesting solutions to a problem.[13] Most leaders across leadership and C&I attributed the source of the problem of poor student performance on state assessments to inadequacies in instruction (rather than, for example, locating the source of the problem in the students or their communities). However, leaders in the two units differed in why they thought instruction was inadequate. Leaders in C&I tended to emphasize that most teachers had not developed sufficiently ambitious instructional practices, such that they could adequately support students' development of conceptual understanding, reasoning, and problem-solving. Meanwhile, leaders in leadership tended to emphasize that teachers were not providing students identified as low performing with adequate opportunities to practice applying procedures to routine problems.

These differences in how leaders across leadership and C&I characterized the source of the problem of poor student performance led to differences in the solutions they proposed to address the problem. Leaders in C&I usually suggested that the fundamental character of instruction needed to change—teachers needed to develop more ambitious and equitable forms of practice. Furthermore, they argued that fundamental changes in instruction would require substantial, sustained learning opportunities for teachers. In contrast, principal supervisors tended to suggest solutions that entailed providing students identified as low performing with additional instruction aimed at improving their ability to solve procedurally oriented test items.

In accounting for these findings, we found it useful to distinguish between two broad orientations that appear to shape how central office leaders frame the problem of advancing student learning, which we term an *instructional improvement orientation* and an *instructional management orientation*. An instructional improvement orientation reflects the view that advancing student learning is fundamentally a problem of improving the quality of instruction. Furthermore, improving the quality of instruction requires substantial learning on the part of teachers in order to support all students' attainment of rigorous learning goals, and thus has implications for coaches' and school leaders' practices. On the other hand, an instructional management orientation reflects the view that advancing student learning requires redeploying the district's existing instructional resources (e.g., human resources, instructional materials). For example, it was not unusual across our partner schools and districts, especially as the time for state testing grew near, for principals to ask coaches to identify students who had struggled on particular state standards, and to identify teachers whose students had performed well on those same standards. The principal would then re-organize the school schedule to match struggling students with the teachers who could work with them on particular standards. As the strategy of matching students and teachers illustrates, there is little reason from an instructional management orientation to support teachers (or coaches or principals) to develop new capabilities.

Our analysis of interviews conducted with central office leaders across all districts in all years of the study indicate that the talk and actions of leaders in C&I tended to reflect an instructional improvement orientation, whereas the talk and actions of central office leaders in leadership tended to reflect an instructional management orientation.[14]

Learning Demands for Central Office Leaders

Against this background, we investigated what central office leaders, especially in C&I and leadership, needed to know and do to design and implement improvement strategies. One aspect of central office leaders' knowledge we conjectured might shape their design and implementation of improvement strategies is their vision of high-quality instruction. We anticipated that leaders in C&I would have developed more sophisticated visions of instruction than those in leadership. However, we found through our interviews that the visions of high-quality mathematics teaching articulated by those in Leadership and C&I did not differ dramatically. Most central office leaders—with the exception of the secondary mathematics leaders and specialists—articulated what we called form-level visions of ambitious and equitable teaching. That is, we found that most central office leaders described several features of inquiry-oriented math instruction that they thought should be evident in classrooms (e.g., real world tasks, teacher questioning, use of manipulatives). However, few central office leaders articulated the intended function of these features in terms of supporting students' development of conceptual understanding of core mathematical ideas.[15] For example, leaders tended to stress the importance of teachers asking students questions, stating that questions are important for getting all students engaged in the lesson at hand. While student engagement is clearly important, leaders did not articulate that a critical function of teacher questioning is to elicit student thinking, thus enabling teachers to make instructional decisions on the basis of students' current understandings. Our finding that most central office leaders (including in C&I, with the noted exception of the mathematics department) articulated form-level visions of ambitious and equitable instruction makes sense—there is little reason why central office leaders would have developed a more sophisticated vision unless they happen to have been mathematics teachers or coaches.

Similar to our findings for school leaders (see chapter 12), we found that it suffices for central office leaders (aside from members of the mathematics department) to share a form-level vision of high-quality instruction as long as they recognize and capitalize on the expertise of colleagues who are capable of supporting mathematics teachers' learning.[16] We found this to be true for senior leaders, like the superintendent and the CAO, as well as for leaders in C&I and leadership. For example, in one of our partner districts, the CAO explicitly indicated he did not have a particularly sophisticated vision of high-quality mathematics instruction. He appreciated the value of engaging students in cognitively demanding tasks and in discussions of their solutions to tasks because he viewed both as essential for supporting students' development of problem-solving and communication capabilities However, he could not describe the function of either aspect of instruction in advancing student learning and did not, for example, distinguish between a show-and-tell discussion and one that advances students' learning (see chapter 3 for a discussion of the distinctions between these forms of discussions.) He, perhaps more than any of the senior leaders we worked with, actively pressed both leadership and C&I to develop compatible agendas that were organized around an ambitious and equitable vision of instruction. Moreover, he almost invariably drew on the expertise of district mathematics specialists, who were somewhat buried within the structure of central office, when formulating policies that were specific to advancing student learning in mathematics. The CAO recognized that he did not have the specific knowledge to make decisions regarding, for example, the design of teacher professional development, and drew on the expertise of those who did.

This example begs the question: Why would someone with a form-level view of high-quality instruction recognize and draw on the expertise of leaders with sophisticated visions of instruction? Certainly, we documented many cases in which leaders articulated form-level views of high-quality instruction and did not draw on the expertise of others. Our analyses of central office leaders, including in leadership and C&I, suggest that leaders with form-level visions of instruction are likely to recognize and seek out colleagues with expertise in mathematics teaching and teacher learning if they approach instructional improvement from a learning perspective, as opposed to a compliance perspective.[17] As David Cohen, an eminent scholar of educational policy and instructional improvement, observed, any strategy focused on advancing student learning that does not simply endorse current practice requires learning on the part of the implementers.[18] A central office leader who embraces a learning perspective, as opposed to a compliance perspective, recognizes that achieving an ambitious and equitable vision of instruction across classrooms and schools is not merely a matter of ensuring compliance with district policies; it requires significant learning on the part of teachers, coaches, and school leaders.

District- and School-Level Decision Making

Thus far, we have focused primarily on challenges that stem from conflicting agendas of the offices C&I and leadership, particularly for the establishment of coherent instructional systems at the school level. Our discussion to this point has assumed that the authority to set direction for instructional improvement efforts resides, for the most part, with central office leaders. Although this was the case for our partner districts for most of the years that we worked with them, it was not invariably the case. As we elaborate below, we found that when decisions about the goals for instruction and instructional improvement were delegated to the school level, it was even more important that school leaders identify and capitalize on the expertise of colleagues capable of supporting the development of coherent instructional systems.

For example, one of our partner districts initially adopted and supported *Connected Mathematics Project 2* district wide. These instructional materials emphasize problem solving and mathematical reasoning and are aligned with an ambitious agenda for mathematics teaching and learning. By the fourth year of our collaboration with the district, a high proportion of teachers used these tasks as the basis for their instruction. As is discussed in chapter 3, the extent to which teachers maintained the cognitive demand, or challenge, of tasks when they implemented them in their classrooms varied considerably. Nonetheless, increases in the rigor of the tasks that teachers use can be considered a significant improvement in the quality of mathematics instruction, even when there is variation in implementation. Prior research indicates that posing rigorous tasks provides students with richer opportunities to learn mathematics.[19] Further, as we reported in chapter 9, the findings of a MIST analysis indicate that the use of cognitively demanding tasks, even when their level of rigor was reduced over the course of the lesson, appeared to positively impact student performance on standardized state assessments.[20]

In year five of MIST, there was a change in this district's senior leadership and the new superintendent and CAO greatly reduced the role of the central office in instructionally focused decision making. The CAO made it clear that central office leaders in C&I could suggest but not require that schools use particular instructional materials. Few principals specified particular materials for their schools and instead left this decision to teachers, with many pulling together materials they found on the internet. As we report in chapter 9, within two years the proportion of teachers using cognitively demanding tasks as the basis for their instruction dropped. Furthermore, our analysis of video recordings of instruction indicated that the rigor of lessons declined substantially once decisions about instructional materials were delegated to schools and teachers. Consistent with this finding, district mathematics specialists reported in annual interviews that the materials teachers were now using lacked coherence and rigor. We found that the school leaders and teachers in most of our partner schools did not have the expertise needed to assemble coherent sequences of mathematics lessons that could support the development of students' understanding of key mathematical ideas. More generally, this example illustrates the detrimental consequences that can ensue when major curricular decisions are delegated to schools *and* school personnel do not have the requisite expertise to select and implement high-quality instructional materials.[21]

KEY TAKEAWAYS FOR CENTRAL OFFICE LEADERS

In light of the findings discussed above, we highlight takeaways specific to supporting leaders across C&I and leadership to establish a shared instructional improvement agenda that aims at rigorous learning goals and an ambitious and equitable vision of instruction. Establishing a shared agenda is crucial if central office leadership is to support the development of coherent instructional systems at the school level.

It is essential that senior district leaders, such as the superintendent and the CAO, set direction for the instructional improvement effort.[22] A key aspect of setting direction entails explicitly communicating to leaders of leadership and C&I that the district is aiming at rigorous learning goals for students and a vision of ambitious and equitable instruction, and that these goals and vision should anchor the design and implementation of instructional improvement strategies. Moreover, it is critical that senior leaders take a learning perspective, as opposed to a compliance perspective. Leaders who take a learning perspective recognize that achieving an ambitious and equitable vision of mathematics instruction across district classrooms is not merely a matter of ensuring compliance with district policies, but instead requires substantial learning on the part of teachers, coaches, school leaders, and district leaders. This implies senior leaders need to do more than just tell leaders in C&I and leadership that they should prioritize rigorous goals and an ambitious instructional vision. They also need to ensure that leaders, especially in leadership and C&I, receive support that focuses on the importance of rigorous goals for all students including those from historically marginalized groups, and on why ambitious and equitable teaching practices are necessary if students are to attain the rigorous learning goals. Moreover, it is important that this support acknowledges and addresses the tension between improving student achievement in the short term and improving the overall quality of instruction in the long term.

We piloted a possible form of support by co-designing and co-leading a one and a half day Instructional Improvement Institute in one of our partner districts with the CAO and secondary mathematics leader. Approximately twenty district leaders from a number of units participated in the Institute, including leaders from C&I and leadership. Goals included coming to appreciate the importance of rigorous learning goals for all students, and developing a shared agenda for instructional improvement for the upcoming year. We first engaged the leaders in activities designed to support them in understanding the capabilities students would need to develop to succeed on new, more rigorous state assessments that assessed problem-solving and conceptual understanding, as well as procedural competencies.[23] We then engaged the leaders in work designed to support them in understanding that achieving these goals would require substantial changes in the quality of instruction in most middle-grades mathematics classrooms—and that realizing those changes (e.g., improving whole class discussion) would require significant teacher learning. Against this background, the leaders were then in a position to begin designing and planning the implementation of instructional improvement strategies for the upcoming year.

More generally, it is essential that leaders across units frame the problem of advancing student learning (especially with respect to rigorous learning goals and increasingly rigorous assessments) from an instructional improvement orientation. That is, leaders need to view advancing learning as a matter of improving the quality of instruction—and recognize that this in turn requires that teachers, coaches, and principals have ongoing professional learning opportunities. At the same time, it is important to acknowledge that improving the overall quality of instruction is a long-term process. A research synthesis published by the National Staff Development

Council suggests that teachers need between 30–100 hours of professional development for 6–12 months to develop inquiry-oriented practices that result in noticeable improvements in student achievement.[24] This means that especially in the initial years of the improvement effort, it is unlikely that most teachers will develop ambitious and equitable teaching practices quickly enough to enable all their current students to attain rigorous goals. These students will therefore be unlikely to receive the high-quality learning opportunities that they deserve.

Given the time it takes for teachers to develop ambitious and equitable instructional practices, there is a role for improvement strategies that reflect an instructional management orientation as well as for those that reflect an instructional improvement orientation, especially in addressing the persistent problem of supporting students identified as low-performing. However, both instructional improvement and instructional management strategies should aim at rigorous student learning goals. Designing instructional management strategies that target rigorous learning goals in the short term will, we imagine, require learning on the part of most leaders in both C&I and leadership. In fact, we did not observe a single case in which either central office leaders or school leaders implemented a strategy reflective of an instructional management orientation that aimed at rigorous learning goals—but we think it is possible. For example, we can imagine school leaders implementing a version of the "matching students with teacher strategy" described earlier, but with a focus on supporting students' attainment of rigorous learning goals, rather than using procedures to solve routine problems. In this case, students who need additional support on specific standards would be matched with teachers who have demonstrated expertise in enacting ambitious and equitable instructional practices and whose students performed well on those specific standards. The school leader could also arrange the schedule such that the "matched classroom" becomes a focus of professional learning for teachers who were less successful in supporting students to attain those standards. For example, teachers might observe the expert teacher's instruction with an accomplished coach, and then discuss their observations in a teacher collaborative meeting. They might then try out some of the observed practices in their own classrooms with the support of the coach. We emphasize that this strategy aims to support students in attaining specific rigorous learning goals in the short term while also contributing to long-term improvements in the overall quality of instruction, *without compromising a commitment to supporting all students to develop rigorous understandings of mathematics.*

Once leaders across leadership and C&I have established common learning goals and a shared instructional vision, and have framed the problem of advancing student learning in similar ways, they are well positioned to collaborate productively on the design and implementation of instructional improvement strategies. It is essential that leaders in leadership substantially involve colleagues with both content-specific expertise and expertise in supporting professional learning (e.g., district mathematics specialists) in the development of content-specific instructional improvement strategies. For their part, it is essential that leaders in C&I collaborate with their colleagues in leadership so that they ensure that the professional learning strategies they design take into account the realities of the various school settings in which teachers teach. We have emphasized repeatedly the importance of ensuring that the various supports for instructional improvement are aligned and aimed at an explicit set of rigorous goals for students' learning and an ambitious and equitable vision of high-quality instruction. This alignment is only possible if leaders of key central office units regularly communicate and collaborate around the work of improvement.

We recognize that regular communication and collaboration between different departments in a district can be a challenge. We found that while leaders across leadership and C&I often intended to collaborate on a regular basis, the multiple, competing demands on their time made it difficult for them to do so. Our experience suggests that it is essential that senior leaders organize leaders' roles and responsibilities so that they collaborate as part of their daily work. For example, in one of our partner districts, the CAO and superintendent re-organized C&I and leadership with the goal of increasing collaboration between the two offices. In the new organization, the district-based math coaches each reported to a principal supervisor, with the expectation that the supervisors could now more regularly and easily draw on the expertise of the coaches in both designing and implementing instructional improvement strategies (e.g., deciding schools on which to focus and how best to support them). This was the same CAO with whom we co-planned and co-led the Instructional Improvement Institute described above in which leaders across units developed instructional improvement strategies. The reorganization of role groups allowed leaders across units to more easily collaborate on the strategies they initially designed.

Last, we specifically address cases in which instructional decision-making resides primarily at the school level. In previous chapters, we described the challenges that school leaders encountered in implementing various elements of a coherent instructional system, including teacher collaborative meetings, instructional coaching, and supplemental supports for currently struggling students. In addition, in chapter 9 we described the challenges in supporting teachers' development of ambitious and equitable practices when they do not have access to rigorous instructional materials. In light of these challenges, we suggest that in districts in which instructional decision-making resides primarily at the school level, district leaders should ensure that schools have access to individuals with expertise in supporting teachers' development of ambitious and equitable instructional practices and in selecting high-quality materials. In our view, the role of these district leaders should be primarily educative and aim to support the learning of school personnel about, for example, the capabilities that students need to develop to perform well on increasingly rigorous state assessments, and the implications for both classroom instruction and for the types of support that teachers require to improve the quality of their instruction.[25] These district leaders would view the development of school-level capacity for instructional improvement as one of their primary responsibilities.

FUTURE RESEARCH

Our work with the four partner districts confirms that central office leaders' expectations and practices influence the extent to which coherent instructional systems are developed and maintained at the school level.[26] However, we also found that the existing research was rather limited regarding what, specifically, central office leaders needed to *know and do* in order to support the development of coherent instructional systems, especially systems organized around rigorous goals for students' learning and a vision of ambitious and equitable instruction.[27] As an example of this kind of scholarship, we have appreciated the research of Meredith Honig, who has worked to identify the *practices* of principal supervisors, and their influence on what principals do in organizing their schools for instructional improvement.[28] We see a pressing need for additional research of this kind that focuses on the knowledge, perspectives, and practices of accomplished senior leaders, as well as leaders in leadership and C&I. Relevant questions that are important to pursue include: What can senior leaders do, in practice, to ensure that leadership and C&I are on the same page and can collaborate productively? What do members of the leadership department need to know to support principals to develop school-wide capacity for instructional improvement? How can principal supervisors support principals to draw on the expertise of C&I in making curricular, professional learning, and human resource decisions?

Additional studies of accomplished leadership practice would serve to clarify goals for central office leaders' learning. There is also a pressing need for research focused on how central office leaders can be supported to develop the necessary forms of knowledge, perspectives, and practices. For example, what forms of professional learning can support leaders across C&I and leadership to design and implement improvement strategies that address both improving achievement in the short-term and improving instruction in the long-term, without compromising a commitment to rigorous learning goals? We described an Instructional Improvement Institute above that might serve as the basis for a feasible kind of support. The field would benefit from studies of such supports for professional learning, in relation to the development of central office leaders' knowledge, perspectives, and practice.

FOURTEEN

Assessing the Impact of Partnership Recommendations on District Instructional Improvement Strategies

ERIN HENRICK, AMANDA B. KLAFEHN, AND PAUL COBB

AS WE DESCRIBED in chapter 2, our pragmatic goal was to add value to our partner districts' instructional improvement initiatives. Our plan for doing this was straightforward. Each year, we documented the districts' current instructional improvement strategies, assessed how these strategies were actually being implemented in schools and classrooms, and shared our findings and recommendations about how the strategies might be revised to make them more effective. However, as our role was purely advisory, it was unclear whether district leaders would use the findings and recommendations when they developed improvement plans for the next school year. (See table 14.1 for examples of recommendations.)

It is important to stress that the support we provided the districts was relatively modest. Although we hoped that our findings and recommendations would influence the districts' improvement strategies, the central office leaders were primarily responsible for implementing those strategies. We did provide some support in the final four years when we worked

TABLE 14.1 Sample recommendations from District Feedback and Recommendation Reports

SAMPLE RECOMMENDATIONS FROM DISTRICT FEEDBACK AND RECOMMENDATION REPORTS
Carefully consider the coherence of the middle-grades mathematics curriculum, especially given the need to condense CMP2 into two grades.
Coordinate and ensure the focus of district- and school-based teacher professional development (PD) remains almost exclusively on high-quality CMP2 implementation.
Principals, together with instructional coaches, develop a clear understanding of the district's vision of high-quality mathematics instruction by participating in math-specific PD.
All instructional coaches should have the opportunity to participate in coaching PD on the curriculum and coaching in a sustained manner as part of the monthly coaching meetings.
The district should continue to provide PD that focuses on high-leverage instructional practices in mathematics and is aligned across role groups (school leaders, coaches, and teachers). The individuals leading PD next year should receive more support and training.
The content of district-wide math PD should focus on specific teacher routines for the whole class discussion in the context of the CMP2 curriculum.
Identify teachers with sufficient expertise and leadership potential, and support their development as formal or informal instructional leaders. Provide qualified teacher leaders a release period for co-planning, co-teaching, modeling instruction, observation, and post-observation conferences with colleagues.
The district should coordinate the curriculum for teacher PD activities across both the district and school levels. In order to successfully coordinate district- and school-based teacher PD, district leaders, school leaders, and instructional coaches need to be on the same page regarding how their training and support aligns and can be made coherent for teachers across the district.

with two districts. First, we led summer planning retreats of a day and a half in which we worked with district leaders to co-design instructional improvement plans for the upcoming year. Second, we collaborated with mathematics leaders in both districts to co-design and in some cases co-lead professional development sessions for instructional coaches and principals.

In order to better understand the extent to which our findings and recommendations added value to our four partner districts' instructional improvement efforts, we compared the recommendations we gave districts over the eight years of the MIST project with the districts' subsequent improvement strategies, and with how those strategies were implemented. In addition, we sought to understand *why some MIST recommendations were taken-up and others were not, and why some recommendations were implemented well and other were not.* Our goal in addressing these questions was to inform the work of other research-practice partnerships that seek to positively impact district improvement efforts.

HOW WE STUDIED TAKE UP OF PARTNERSHIP RECOMMENDATIONS

We determined whether or not our partner districts acted on, or took up, our recommendations by examining the District Design Documents for the following year that describe the districts' improvement strategies for that year. We classified recommendations as "taken up," "somewhat taken up," or "not taken up." We coded a recommendation as "taken up" if the revised strategy appeared in the district's improvement plan the subsequent year. A recommendation was coded as "somewhat taken up" if some aspects of the recommendation were included in the district's improvement plan for the following year. For example, the recommendation that a district provide both a mathematics-specific form for school leaders to use when they observed instruction and professional development on how to use the form was coded as somewhat taken up if the district provided such a form but did not provide professional development. We coded a recommendation as "not taken-up" if there was no mention of it in the District Design Document for the following year. We then looked for themes related to why some recommendations were taken up and others were not.

FINDINGS REGARDING TAKE-UP OF RECOMMENDATIONS

Over the course of the study, we made 162 separate recommendations. Looking across all of the district and years, we found that the districts attempted to act upon 67 percent of the recommendations, meaning the recommended strategy appeared in the district's improvement plan for the subsequent year. All the districts attempted to act upon at least 50 percent of our recommendations, but take-up ranged from a low of 56 percent in one district to 82 percent in another. These findings are encouraging given that we provided that districts with only limited support (primarily the annual feedback reports and meetings) and the results that we shared with district leaders were frequently disappointing for them. We conjecture that district leaders acted on a relatively high proportion of our recommendations because the analyses we reported focused specifically on the districts' improvement strategies and thus related directly to the work district leaders were attempting to accomplish.

When we sought to understand why some recommendations were taken-up and others were not, we found that take-up did *not* vary by either the role group on which the recommendation focused (i.e., teachers, coaches, school leaders, district leaders) or by the goal of the recommendation (i.e., what the strategy aimed to achieve.) Instead, we found that recommendations were more likely to be taken up when they aligned with district priorities, when district leaders who were central to the partnership could authorize their implementation, and when district leaders could devote adequate expertise and resources to implementation.

The Alignment of Recommendations with District Priorities

Recommendations that aligned with a district's priorities were more likely to be taken up. For example, a district took up only one of seven recommendations one year because the district's priorities had shifted away from mathematics teachers' development of ambitious and equitable instructional practices due to a change in senior district leadership. Although it makes sense that our recommendations were not taken up, this example also highlights the challenges of a research team and a district maintaining a common improvement agenda when district leadership is in flux.

Recommendations that built on ongoing mathematics initiatives were more likely to be taken up. We recommended to one district that the district mathematics department ensure that district- and school-based teacher professional development be organized around the adopted instructional materials, *Connected Mathematics Project 2* (CMP2). Because the district mathematics department already aimed to support the successful implementation of the CMP2 materials, this recommendation was taken up. The following school year, the district mathematics specialists designed and led district-level and school-level professional development that focused on using upcoming CMP2 units.

Recommendations were also more likely to be taken up if they built directly on current activities and did not require ongoing support to implement. For example, in one district, we recommended that school leaders periodically attend professional development for mathematics teachers in order to better understand mathematics instruction. We conjecture that this recommendation was taken up because the teacher professional development activities were already taking place, and this recommendation could be addressed with a district directive.

Conversely, we found that recommendations were less likely to be taken up if they competed for resources with district initiatives that cut across content areas. For example, we recommended to one district that that professional development for mathematics teachers focus on the structure of CMP2 lessons (i.e., launch, explore, summarize). This recommendation was not taken up because the district made formative assessment in all core content areas a priority the following year, and provided professional development with this focus.

District Leaders' Role in Taking Up the Recommendation

Recommendations were more likely to be taken up when district leaders central to the partnership team could authorize their implementation. For example, in one district, we recommended that the district hire additional district-level instructional coaches. Because we worked closely with district personnel who had budget and hiring authority, this recommendation was taken up. Across the districts, most of the recommendations that pertained to the work of members of the mathematics department were taken-up because district mathematics leaders usually had the authority to act on our recommendations and saw them as contributing to the attainment of their improvement goals. Additionally, the recommendations that involved district leaders and members of the research team co-designing and co-implementing initiatives were all taken up. For example, we recommended to two districts that they conduct professional development with school leaders on the lesson structure of a CMP2 lesson. As a part of our collaborative work the following year, we co-designed and co-led such activities. These examples illustrate the potential impact of ongoing collaborations between researchers and district leaders.

Conversely, recommendations typically did not get taken up when they required coordination across multiple central office departments. For example, recommendations that required coordinated decision-making between multiple district offices (e.g., the district unit that supported principals and the district mathematics department) were usually not taken up. We found that when cross-department recommendations were taken up, district leaders with the authority to address the recommendation had attended the District Feedback and Recommendations Report (DFRR) meeting. We conjecture that participation in the meeting supported the development of a shared understanding of the rationale for the recommendation. We also found that the influence of a "partnership champion" was relevant to cross-department recommendations that were taken up. A greater proportion of cross-departmental recommendations were taken up in the districts where one or more senior district leaders was central to the work of the research-practice partnership (e.g., superintendent, assistant superintendent, chief academic officer). In these cases, a senior district leader valued the work of the partnership, understood the connection between improving the quality of instruction and increasing student achievement, and played a critical role in decision-making that spanned across central office departments.

It was also the case that recommendations frequently did not get taken up when they required coordination across multiple role groups. For example, recommendations that required principals and district leaders to work together to establish a shared vision of high-quality mathematics instruction were usually not taken up. This was frequently because the recommendations did not fall under a single person's job responsibilities and it was not clear who should take responsibility for implementing the recommendation.

Expertise and Time Needed to Take Up Recommendations

Recommendations that required significant expertise and resources were less likely to be taken up. Even though the research team tried to make recommendations that were feasible given the districts' current capacities, there were instances where the districts did not have the relevant expertise or resources necessary to take up the recommendation. For example, on several occasions, we recommended that instructional coaches receive professional development on one-on-one coaching practices that that have the potential to support teachers' learning (e.g., modeling instruction, co-teaching, and enacting the coaching cycle). These recommendations were usually not taken up due to limited expertise within the district in supporting coaches' learning and lack of available funds for hiring external professional development providers. We did find that recommendations that required significant expertise and resources were more likely to be taken up when a senior district leader with broad decision-making responsibilities and vision for improvement that aligned with that of the research team championed the partnership.

Conversely, we found that recommendations were often taken up if they required limited expertise and resources. For example, recommendations that could be implemented by making a district level directive (e.g., all schools should schedule time for teacher collaborative meetings) were typically taken up provided they aligned with current district priorities *and* the district leaders with decision-making authority were actively involved in the MIST partnership. These recommendations usually involved making changes at the form but not the function level (e.g., scheduling time of teacher collaborative meetings but not supporting improvements in the quality of the meetings).

HOW WE STUDIED HOW RECOMMENDATIONS WERE IMPLEMENTED

We also investigated what happened when districts attempted to implement MIST recommendations. In this second phase of the analysis, we aimed to answer the questions: To what extent did the districts implement the recommendations that they took up successfully? Why were some recommendations implemented successfully while others were not? We assessed whether the implementation of each recommendation that was partially or completely taken up positively impacted the districts' improvement efforts. We made these assessments by drawing on the District Feedback and Recommendations Reports for the following year in which we documented how the districts' improvement strategies were being implemented. We coded each recommendation as being either successfully implemented, partially implemented, or unsuccessfully implemented. The criteria for successful implementation were that there was evidence that the recommendation was fully implemented the following year *and* there was evidence that it positively impacted the district's improvement effort. A recommendation was coded as somewhat successfully implemented if there was evidence that the recommendation was either fully implemented or partially implemented *and* there was evidence that this implementation had a moderately positive impact on the district's improvement efforts. Recommendations were coded as not successfully implemented if the recommendation was fully implemented or partially implemented, *but* there was little to no evidence of a positive impact on the improvement efforts.

FINDINGS REGARDING IMPLEMENTATION OF RECOMMENDATIONS

As reported above, the districts acted on 67 percent of our recommendations. However, of the 109 recommendations that the districts acted upon, we coded only 17 percent of the recommendations as successfully implemented, 27 percent as somewhat successfully implemented, and 56 percent as minimally or unsuccessfully implemented. These findings make it clear that while it is important for district leaders to draw on research when making decisions about improvement strategies, this is by no means sufficient. As these findings and the findings reported in prior chapters indicate, the districts often encountered challenges when they attempted to implement our recommendations. As we have noted, we could provide the districts with only minimal support for implementation. The findings therefore indicate the importance of making the implementation of potentially productive improvement strategies an explicit focus of investigation. We take up this issue in the chapter 15 when we discuss the implications of the work presented throughout the book for future research.

Our analysis of why some recommendations were implemented successfully while others were not indicated that success was not related to either the role group on which recommendations focused (e.g., teachers, coaches, school leaders, district leaders) or by the goal of the recommendation (e.g., what the strategy aimed to achieve.) Instead, we found that recommendations that were successfully implemented were under the purview of district leaders centrally involved in the partnership and that the district personnel responsible for the implementation had the requisite expertise.

District Leaders' Role in Implementing the Recommendation

Two-thirds of the recommendations that were successfully implemented were implemented by the district mathematics department. For example, in one district, the state mandated that all ninth-grade students should take algebra I. We recommended that in responding to this mandate, the district math department ensure that the middle-grades mathematics curriculum remained coherent, especially given that the adopted curriculum (CMP2) could only be used in two grades rather than three as intended. This recommendation was successfully implemented because the math department has the necessary time and expertise to develop a document that aligned CMP2 with the new state standards. This document supported middle-grades mathematics teachers by specifying what to teach, indicating central mathematical ideas, identifying gaps between CMP2 and the new standards, and indicating instructional resources that could be used to address these gaps.

One third of the recommendations that were successfully implemented were under the purview of a senior district leader who was actively involved in the partnership. For example, we recommended to one district that its monthly principal meetings focus regularly on the district's vision for high-quality mathematics instruction and how the adopted instructional materials support this vision. The associate superintendent in charge of school leadership played a key role in the partnership and she made this a component of the principal meetings, drawing on the expertise of district mathematics leaders to develop and lead this professional development. Principals reported in the interviews that we conducted with them that the principal meetings regularly focused on mathematics teaching, and we also found evidence that their visions of high-quality mathematics instruction had improved compared with prior years. The following year, we recommended to the same district that all its instructional leadership initiatives emphasize key aspects of high-quality mathematics instruction. That year the members of the research team co-designed and co-led three professional development sessions for principals and coaches with members of the district mathematics department. We found that principals' visions of high-quality mathematics instruction had further improved and were more closely aligned with the district's vision.

Expertise and Time Needed to Implement the Recommendation

We found that when recommendations were implemented successfully, the district personnel responsible for the implementation had the necessary expertise. For example, we recommended to one district that the mathematics department provide professional development for the school leaders in charge of mathematics at each school, with a focus on what high-quality implementation of rigorous tasks in the upcoming curriculum units should look like. The district mathematics specialists who designed and led this professional development had expertise in mathematics instruction and a deep understanding of the adopted instructional materials. We subsequently found that the participating school leaders' visions of high-quality mathematics instruction had improved.

Conversely, although we aimed to make recommendations that were feasible, we found that when recommendations were not implemented successfully, it was frequently because the district personnel charged with implementation did not have the necessary expertise. For example, in order to successfully implement recommendations related to improving the quality of teacher professional development, it was important that the members of the mathematics department had expertise not just in ambitious and equitable mathematics instruction, but also in supporting teachers' learning and in designing and facilitating professional development.

An additional common reason why many recommendations were not implemented successfully related to competing priorities and limited time. For example, recommendations related to teacher and school leader professional development were frequently not implemented successfully due to low attendance, which in turn was due to difficulties in scheduling the professional development sessions when all the intended participants could attend. Further, recommendations related to coach professional development were often not implemented successfully because professional development sessions had to be cancelled or rescheduled due to other competing priorities and pressing needs.

TAKEAWAYS FOR RESEARCHERS

The finding that district leaders acted upon two thirds of the research-based recommendations is encouraging and goes some way to countering frequent reports of practitioners failing to take account of research findings. In addition, the finding reported in this chapter that recommendations were more likely to be taken up when they aligned with ongoing district priorities indicates that the approach of organizing our findings and recommendations around the districts' current improvement strategies contributed to the high uptake. The analyses reported in this chapter also clarifies that uptake was more likely when district leaders who were central to the partnership could authorize implementation, and when district leaders could devote adequate expertise and resources to implement the recommendation.

In contrast to the encouraging findings about the uptake of recommendations, we found that our partner districts frequently encountered difficulties when they attempted to act on our recommendations. We found that most of the recommendations that were successfully implemented were under the purview of district leaders centrally involved in the partnership and that the district personnel responsible for the implementation had the requisite expertise.

These findings have implications for the work of researchers aiming to influence practice. First, it appears critical that practitioners view findings as addressing their current problems of practice and thus as directly relevant to their ongoing work. This points to the importance of ensuring research plans take account of practitioners' ongoing priorities and initiatives.[1]

Second, our findings indicate the value of teams including practitioners with relevant expertise and with the authority to facilitate the take-up and implementation of findings and recommendations. For example, even though the focus of our district partnerships was on middle-grades mathematics, the involvement of district personnel in other divisions in partnership work contributed to the uptake of recommendations and to successful implementation of a greater proportion of the recommendations than would otherwise have been the case.

Third, our finding that our partner districts encountered difficulties when they attempted to implement many of our recommendations indicates the value of future studies that aim not only to identify potentially productive improvement strategies but also to investigate and support the implementation of those strategies. Our findings indicate that studies of this type should develop conjectures about the personnel, types of expertise, tools, and time that the successful implementation of particular strategies might require, and then test and revise those conjectures while investigating how the implementation of the strategies is playing out in a range of different contexts. We return to this issue in chapter 15 when we discuss the implications of the findings reported throughout this book for research.

FIFTEEN

Putting the Pieces Together

PAUL COBB, KARA JACKSON, ERIN HENRICK,
AND THOMAS M. SMITH

IN THE OPENING CHAPTER, we shared the story of the NUMMI car plant. Our purpose was not to suggest that improving the quality of teaching is directly analogous to assembling cars reliably, but instead to clarify the value of taking a *systems perspective*. In the NUMMI case, the system for building high-quality cars spanned from the shop floor to the head office. Similarly, the Theory of Action (ToA) we have proposed for supporting instructional improvement on a large scale is broad in scope and spans from the classroom to the district central office. As we noted, this ToA takes on increasing significance with the implementation of more rigorous state standards and assessments.

In the subsequent chapters, we reported findings and shared actionable takeaways for each component of the ToA for instructional improvement at scale, and discussed the implications for future research. Throughout, we pointed to the connections between the components so that the reader might get a sense of how the various pieces of the ToA fit together and constitute a system. In the first part of this chapter, we step back to focus on issues that cut across the components. We then discuss the implications of the work we have reported for the practice of instructional improvement and for research.

CROSS CUTTING ISSUES

Systems Perspective

As we noted in chapter 1, student learning goals and thus what counts as high-quality instruction are at the core of the ToA. In specifying a research-based vision of high-quality instruction in chapter 3, we embraced the challenge facing our partner districts, that of supporting diverse groups of students in attaining rigorous learning goals. Consequently, we attended explicitly to equity in students' learning opportunities when detailing a vision of high-quality mathematics instruction. Throughout, we referred to this vision as "ambitious and equitable."

This vision of high-quality instruction specifies goals for teachers' improvement of their instructional practices and thus for teachers' learning. It was therefore essential that the remaining elements of the coherent instructional system were aligned with this vision if they are to contribute to teachers' development of instructional practices that enable *all* students to attain rigorous learning goals. With this in mind, we followed a backward mapping approach by putting students' learning at the center of improvement work and delineating a coherent system of supports for teachers' learning (see figure 15.1). This system in turn has implications for the practices of school and district leaders, the tools they use, and supports for their learning.

One immediate consequence of this approach is that the student learning goals and instructional vision matter all the way up and have implications for both school and district instructional leadership. Conversely, we illustrated in chapter 13 that the relations between the central office units of leadership and of curriculum and instruction matter all the way down and impact classroom teaching and learning. As a further indication of the systems character of both the ToA and of the reality of instructional improvement, issues of equity are not cordoned off as a separate component, but instead permeate all aspects of the ToA.

We suggest that this approach of first focusing on the classroom by specifying student learning goals and an explicit vision of high-quality instruction, and then mapping out is relevant to schools' and districts'

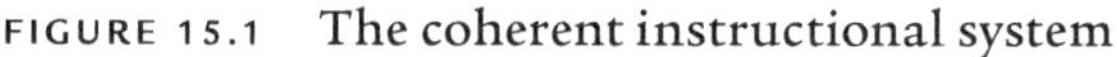

FIGURE 15.1 The coherent instructional system

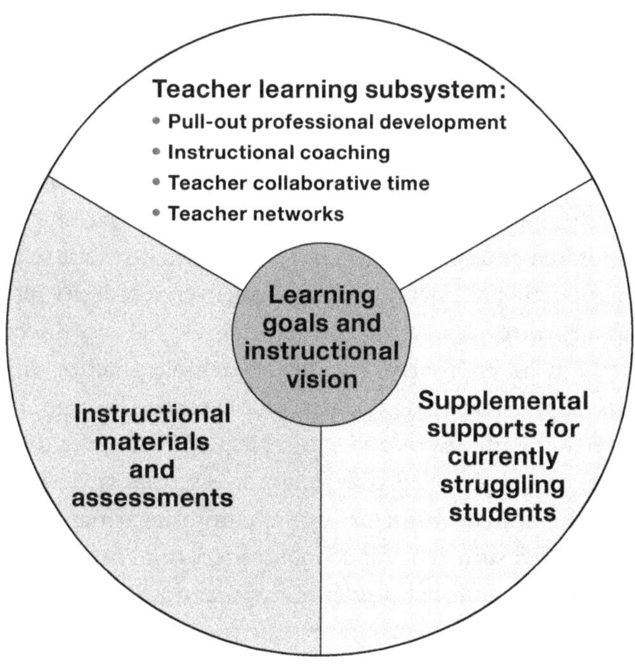

development of instructional improvement plans. Based on our work with the four districts, we doubt that mapping backwards from student learning goals is the standard approach that leaders take when crafting instructional improvement policies. However, the approach of mapping backwards from goals (in the case at hand, goals for students' and teachers' learning) is routine in design professions.[1]

The Importance of Expertise

The critical role of expertise features heavily in many components of the ToA including all facets of the teacher learning subsystem, school leadership, and district leadership. This is not surprising, given the challenging nature of supporting all students to attain rigorous learning goals, and the fact that doing so requires substantial improvements in most teachers' daily practice, as well as in coaches', school leaders', and district leaders' practices.

For example, across the facets of the teacher learning subsystem, we detailed the critical importance of teachers working routinely with an accomplished colleague on improving their practice. As we specified in chapter 7, this involves working with an instructional coach—but it is crucial that a coach has not only developed ambitious and equitable instructional practices but has also developed expertise in supporting other teachers' learning. Similarly, in chapters 5 and 6, we emphasized the importance of ensuring that pull-out professional development and teacher collaborative time are facilitated by a coach or teacher leader with expertise in both teaching and in supporting teachers' learning. And, in chapter 8, we reported that teachers who turn to colleagues who have developed more accomplished practices for advice improved their own teaching practice. Importantly, across these chapters, we also indicated that expertise in supporting teachers' development of ambitious and equitable practices was in short supply.

In chapter 12, we reported that it is not essential that school leaders develop a sophisticated vision of high-quality mathematics instruction. Instead, it appeared sufficient for school leaders to appreciate the value in pursuing rigorous goals for all students and that doing so requires substantial learning for most teachers. However, it is essential that school leaders recognize the expertise of and collaborate with colleagues who are accomplished in supporting teachers' learning (e.g., a district math specialist, an accomplished coach). Similarly, as we reported in chapter 13, the more effective senior leaders (e.g., chief academic officer) with whom we worked capitalized on the expertise of district math specialists when making decisions related to mathematics teaching and learning.

Taken together, these findings emphasize the critical role of accomplished coaches and district math specialists. In the course of our collaborations with the four districts, we worked with a number of different math specialists and saw firsthand the difference that an accomplished specialist can make in supporting both district leaders and school instructional improvement efforts by, for example, designing and leading high-quality pull-out professional development for teachers and for coaches.

We also reported findings that clarify what counts as expertise in teaching and in supporting teachers' learning. For example, in chapter 3 we showed that while mathematical knowledge for teaching is important, the quality of teachers' instruction is unlikely to improve unless they have developed relatively sophisticated visions of high-quality mathematics instruction and productive views of their students' current mathematical capabilities. In chapters 5 and 6, we addressed the question of what facilitators of professional learning need to know and be able to do to engage groups of teachers in inquiring into and improving their teaching practice. Our findings indicate that accomplished facilitators know how to establish professional communities in which teachers share ideas and take intellectual risks, establish clear goals for teachers' learning, and are skilled in pressing on teachers' contributions to achieve a professional learning agenda while maintaining a culture of trust. In chapter 7, we addressed the question of what coaches need to know and be able to do to support teachers' learning beyond being accomplished teachers themselves. Our findings indicate that accomplished coaches have developed what we called a sophisticated professional vision for coaching, meaning that accomplished coaches assess teachers' current classroom practices, locate those practices on a long-term trajectory for teachers' learning, and select the types of activities in which to engage teachers accordingly.

Taken together, our findings point to a distributed model of instructional leadership in which leadership functions, such as establishing an orienting vision for instructional improvement, are carried out by several leaders working in concert.[2] However, as Richard Elmore observed, distributed leadership is likely to be, at best, ineffective unless it also follows what he termed the "contours of expertise" wherein a leader with greater expertise on a particular aspect of the work takes the lead for that aspect.[3] Elmore went on to note that a leadership model that follows the contours of expertise might sound radical in the context of public education but is routine in most knowledge-based enterprises.[4]

A model of instructional leadership that follows the contours of expertise places a premium on school and district leaders in formal positions of authority identifying relevant forms of expertise and recognizing and collaborating with colleagues who have already developed those capabilities. In the course of work with the districts, we saw instances in which the expertise of colleagues who were accomplished in teaching and in supporting teachers' learning repeatedly went unrecognized in both the central office and in school buildings, to the detriment of the district' instructional improvement efforts. These observations serve to emphasize that the ToA does not consist of a list of prescriptions that should be slavishly followed. As we clarify later in this chapter, developing improvement plans that are informed by the ToA involves wisdom and judgment, a key aspect of which involves identifying, recognizing, and capitalizing on relevant forms of expertise.

Influence of State Accountability Systems

In developing the ToA, we attempted to limit our focus to policies and processes that were within the locus of control of schools and districts. As a consequence, we did not set out to investigate the influence of the state accountability systems on our partner districts' instructional improvement efforts. However, we knew that in aiming at rigorous learning goals for all students while the state assessments were procedurally oriented, our partner districts would be flying into a strong headwind.[5] It should therefore be no surprise that we repeatedly found ourselves documenting the influence of these accountability systems on the districts' instructional improvement efforts.

We should stress that the influence of state assessments on classroom instruction was not direct, but was mediated by principals and by district leaders, particularly those who supervised and supported principals. There are repeated indications in the prior chapters that when principals perceived that they were primarily accountable for improving student achievement on procedurally-oriented assessments rather that supporting teachers' development of ambitious and equitable practices, they tended to act in ways that they thought would result in improvements in achievement. For example, many principals implemented second math classes for currently struggling students that focused on procedures rather than on participating effectively in and learning from their mainstream mathematics classes. As we reported in chapter 11, procedurally oriented second classes generally failed to improve students' performance on procedurally oriented state assessments.

In addition, principals who perceived they were primarily held accountable by their supervisors for improving student achievement on procedurally oriented assessments implemented supports for teachers' learning in ways that dramatically limited teachers' opportunities to improve the quality of their teaching. For example, as we described in chapter 7, such principals were likely to assign school-based coaches additional duties related to state accountability testing that significantly reduced the proportion of their time that they spent working with teachers on instructional issues.

These additional duties included tutoring individual students, identifying students for second mathematics classes, locating or creating curricula and materials for second math classes, setting up and managing the student log-in process for computerized test preparation software, and proctoring student assessments.

Furthermore, as we described in chapter 6, principals who focused on improving student achievement were likely to steer teacher collaborative groups to engage in activities intended to boost student achievement such as identifying students for second math classes and determining which standards needed to be retaught. In these cases, teacher collaborative groups did not support teachers in improving the quality of their instruction. In addition, also reported in chapter 8, such leadership practices communicated to teachers that students' performance on state assessments was the bottom line. This in turn impacted teachers' advice networks by influencing teachers to seek advice from colleagues whose students performed well on state assessments rather than from those who had developed more accomplished instructional practice.

It is perhaps not surprising that principals who felt pressure to increase student achievement on procedurally oriented assessments tended to orient teachers to teach (and reteach) procedures. It is, however, unfortunate as the findings of a substantial number of studies consistently indicate that students who receive inquiry-oriented instruction perform as well as students who receive procedurally oriented instruction on assessments of procedures, and better on assessments of conceptual understanding and problem solving (e.g., see chapter 3).[6]

This latter finding takes on increasing significance with the implementation of more rigorous assessments in most states. However, our work with school and district leaders indicates that many will require support if they are to understand the new capabilities that students need to develop in order to perform well on the more rigorous assessments. As we discussed in chapters 12 and 13, it is also important that they understand the implications of the more rigorous learning goals for teachers' instructional practices and thus for the ongoing support that teachers will require to develop such practices. Only then might they recognize the expertise of colleagues who can support teachers' learning and implement improvement strategies aimed at improving the quality of instruction.

IMPLICATIONS FOR THE PRACTICE OF IMPROVEMENT

Developing Instructional Improvement Plans

The ToA that we propose in this book specifies a range of productive improvement strategies and indicates how they are interrelated and constitute a system. It is best viewed as a resource on which school and district leaders can draw when formulating instructional improvement plans that are tailored to their specific contexts. In this regard, the ToA can serve as an empirically grounded touchstone for school and district leaders, who are often bombarded with solicitations from vendors promoting their programs and services.

In addition to proposing specific improvement strategies, we have illustrated a general approach for developing improvement plans that we contend is relevant to school and district leaders. This approach involves first specifying student learning goals and an explicit vision of high-quality instruction that is both ambitious and equitable (see chapter 3). As we noted in chapter 1, these goals and instructional vision are the core of a coherent instructional system. The next step is to map out from the learning goals and instructional vision by designing the remaining three elements of a coherent instructional system. This involves making decisions about instructional materials and assessments, and about associated district-developed resources such as curriculum frameworks, that are aligned with the learning goals and instructional vision (see chapters 9 and 10). It is also essential to design a system of supports for teachers' learning that is organized around the instructional materials (see chapter 4) and to consider supplemental supports for currently struggling students that will enable them to participate effectively in and learn from their mainstream classes (see chapter 11). The final step in developing an instructional improvement plan is to spell out the implications of the above elements of a coherent instructional system for coaches', school leaders', and district leaders' practices, for the tools they will use, and for the supports for their development of those practices (see chapters 7, 12, and 13). The watchwords in this approach are alignment and coherence, with the goal of ensuring that the various improvement strategies work together and constitute a system that can support improvements in the quality of teaching and students' learning.

Improving Instructional Improvement Plans

We developed an analytic framework while working with our partner districts that proved crucial in enabling us to make empirically grounded recommendations to the districts about how they might revise their improvement strategies. In describing the framework in chapter 2, we explained that improvement strategies that have the potential to support consequential professional learning are likely to involve some combination of:

- Ongoing intentional learning events
- On-the-job support and guidance from a more accomplished colleague
- Carefully designed organizational routines carried out with a more accomplished colleague
- New tools together with support on learning to use them effectively

This framework can be of value to school and district leaders as it enables them to assess whether their improvement strategies are likely to be effective before they are implemented, and to revise them accordingly. To illustrate the usefulness of the framework, we share how we used it to assess the potential of a strategy focused on principals' practices that one of the districts implemented during the second year of our partnership.[7] The leaders of this district intended that principals would do the following as instructional leaders to support teachers to improve the quality of instruction: a) observe instruction regularly, focus on how the adopted instructional materials and curriculum frameworks were being implemented, and provide feedback on instruction; b) periodically conduct quick, focused observations of multiple classrooms, sometimes with the coach who served the school, to assess building needs; and c) meet regularly with the coach to ensure that teachers were receiving appropriate support.

The first step in assessing whether this improvement strategy was likely to be effective was to clarify the nature of the principals' learning. We inferred that principals would have to distinguish between strong and weak inquiry-oriented instruction in order to communicate expectations for instructional improvement that were consistent with the district's theory of action when making classroom observations. We also inferred that if principals could distinguish between strong and weak inquiry-oriented instruction, they might be able to identify teachers' needs and, in collaboration with the coach, plan or procure additional supports for their learning (e.g., from district mathematics specialists or external consultants). However, our findings indicated that principals would need significant professional learning before they could distinguish between strong and weak instruction. As described in chapter 2, we asked principals in the interviews we conducted with them each year to describe their vision of high-quality mathematics instruction. As reported in chapter 12, most principals had developed, at best, a form-level view of ambitious and equitable instruction. For example, the principals in this district suggested that teachers should use "real-world problems" and that "students should explain their thinking to one another" but none could distinguish between strong and weak real-world problems, or between discussions that could support students' learning and those that were unlikely to do so.

The second step in assessing the strategy entailed us asking whether the supports for principals' learning were likely to support their development of the intended practices. We were doubtful because the supports did not include *ongoing intentional learning events* that focused on distinguishing between strong and weak inquiry-oriented instruction. The district Mathematics Department had developed curriculum maps for principals to use when making classroom observations that described the mathematical concepts on which instruction was expected to focus, the resources teachers should use, and expected student products (i.e., a *new tool*). However, we doubted that the maps would be sufficient as no supports were planned on using them effectively. We noted that the regular meetings with the coach might provide *on-the-job support and guidance from a more accomplished colleague,* but this would be fortuitous as the meetings focused on supporting teachers' rather than the principal's learning. Similarly, the observations of multiple classrooms that principals were expected to make with a coach could give rise to learning opportunities, but these would also be fortuitous as this *organizational routine* was not designed to support principal's learning. Our analysis of this strategy indicated that its effectiveness depended unduly on learning opportunities that might arise incidentally when working with the coach.

Stepping back from the illustration, notice how the framework allowed us to identify limitations in a specific strategy before it was implemented. School and district leaders who use the framework in this way would also be able to anticipate difficulties and thus revise strategies before implementing them. In this case, our analysis suggests the need to provide more robust supports for principals' development of increasingly sophisticated visions of high-quality mathematics instruction and for their use of the curriculum maps. In addition, if schools and districts can monitor how particular strategies are actually being implemented, the framework can enable them to *explain* any gaps between the strategy as intended and as implemented. These explanations can then inform decisions about how to revise the strategy or to abandon it.

Continuing the above illustration, our findings indicated that there were several gaps between the intended strategy and the strategy as implemented. For example, teachers reported in interviews that although principals were observing instruction and providing feedback regularly, their feedback typically focused on easily observable aspects of instruction (e.g., objectives posted, presence of word walls) and on whether students were engaged in lessons. Consequently, the feedback that teachers received failed to communicate expectations that were consistent with the district's vision of ambitious and equitable instruction. Furthermore, only one of the seven principals reported observing classroom instruction with a mathematics coach. Principals and coaches both reported that the coaches' schedules made it difficult to schedule learning walks. In addition, principals also reported in interviews that they were not using the curriculum maps to guide their observation of classroom practice.

We used the framework to explain these gaps by focusing on how the supports for principals' learning were actually implemented. For example, the finding that principals' feedback was unlikely to support teachers to improve instruction was not surprising, given the nature of the principals' visions of high-quality mathematics instruction and the absence of supports with this focus. The finding that only one of seven principals observed classrooms with the coach was significant as it limited their opportunities to learn from a colleague with greater relevant expertise. The finding that principals were not using the curriculum maps to guide their classroom observations was also not surprising and is attributable at least in part to the lack of support for learning how to use the maps in practice.

This analysis indicated a number of ways in which the strategy might be revised to make it more effective. For example, it seemed essential to provide the principals with ongoing, intentional professional development designed to support their development of more sophisticated visions of high-quality instruction and their use of the curriculum maps. In addition, it appeared important to reorganize coaches' schedules so that they could observe instruction with principals on a regular basis, and to organize these joint observations so that they could be an intentional support for principals' development of more sophisticated visions of instruction.

Using the framework to improve its strategies in this way, a school or district would eschew the all too common practice of repeatedly abandoning a current strategy and switching to a different one when there are indications that a strategy is not leading to the desired results without first attempting to understand why this is the case. In particular, absent inquiring into why a strategy is not leading to the desired results, it is unclear whether the problem is with the design of the strategy or with its implementation, and whether it is feasible to improve either the strategy or its implementation. While working with our four partner districts, it became evident that schools and districts were more likely to make progress if they developed a sound initial improvement plan and then worked to improve the strategies and their implementation, and only abandon a strategy when it was clear that it was fundamentally flawed.

We found that in arguing for and justifying particular improvement strategies, school and district leaders often convinced themselves that the strategies they proposed would be effective. In contrast, the use of the framework both requires and supports a skeptical attitude towards a design for instructional improvement. This attitude is similar to that of teachers who routinely assess their students' learning as an integral part of their instruction and use what they learn to adjust that instruction. Similarly, in using the framework as we have described, school and district leaders can assess and improve strategies both when designing them and when implementing them.

Customizing the Theory of Action to School and District Contexts

In describing the process that we followed to develop actionable recommendations for our partner districts, we noted that we took account of each district's current capacity for instructional improvement (see chapter 2). Similarly, it is essential to customize the theory of action and the improvement strategies to particular school and district contexts. The first step in customizing the theory of action is to assess the current capacity of schools and the central office for instructional improvement. Doing so entails assessing:

- The current levels of accomplishment of teachers, coaches, school leaders, and key district leaders
- The nature and quality of the tools the members of these role groups are using (e.g., curriculum frameworks, classroom observations forms)
- The nature of the relationships between school and district personnel, including:
    - The extent to which teachers trust their peers and coaches sufficiently to share their problems of practices (see chapter 6)
    - The extent to which coaches and principals are collaborating to support teachers in improving the quality of their instruction (see chapter 7)
    - The extent to which district leaders in the departments of Leadership and of Curriculum and Instruction are collaborating to design and implement instructional improvement strategies (see chapter 13)

This initial assessment of a school's or district's current capacity for improvement delineates the starting points for the instructional improvement effort, and thus the first feasible steps.

As an example, one of our partner districts scheduled a class period for mathematics teachers to collaborative every school day. As we reported in chapter 6, the role of a facilitator is crucial in determining whether teacher collaborative meetings are effective in supporting teachers' development of ambitious and equitable instructional practices. However, few of the district's mathematics teachers had made significant progress in developing ambitious and equitable instructional practices. We were therefore not surprised to find that this expensive initiative had not supported any discernible improvements in the quality of instruction. In this case, the improvement strategy, teacher collaboration, exceeded the teachers' current levels of accomplishment. In chapter 7, we suggested that in such cases coaches should facilitate teacher collaborative meetings. However, this recommendation assumes that coaches are accomplished in supporting teachers' learning. If the initial assessment of a school's or district's current capacity for improvement indicates that this is not the case, we would recommend delaying the implementation of teacher collaborative time and focusing first on developing a cadre of accomplished coaches.

As the above example illustrates, an initial assessment of a school's or district's current capacity can inform the sequencing of improvement initiatives. As a related point, it is also important to limit the number of initiatives that impact the members of particular role groups at the same time. As a rule of thumb, it is unreasonable to expect that teachers can improve more than one aspect of their instruction that involves substantial learning at a time. We assume that this rule also applies to coaches', school leaders', and district leaders' improvement of their practices. It is therefore an important constraint that needs to be taken into account when designing school improvement plans.

Adapting the Theory of Action to Other Content Areas

We have suggested that the theory of action can inform instructional improvement efforts in other content areas, if the goal of the improvement work is to support teachers in substantially reorganizing rather than merely elaborating or extending their current practices. However, as we made clear earlier in this chapter, the theory of action is specific not merely to mathematics but to a particular set of mathematical learning goals and associated vision of high-quality mathematics instruction. Therefore, in adapting the theory of action to other content areas, it is essential to identify a set of clear learning goals in that content and to establish a vision of what high-quality instruction in that content area entails. In our view, it is therefore essential to draw on the expertise of specialists in the relevant content area when adapting the theory of action. Further, we suggest that leaders and specialists follow the approach of mapping out from student learning goals that we have described when deciding whether various components of the theory of action need to be modified, and if so, how.

IMPLICATIONS FOR RESEARCH

The preceding chapters include recommendations for needed research on the various components of the theory of action. As MIST was a relatively large project, it might be assumed that a large research team needs to be assembled and substantial external funding secured in order to make a significant contribution to our understanding of what it takes to improve the quality of teaching and learning on a large scale. However, this is not the case. In chapter 5 we described a design study conducted as part of the MIST project that sought to clarify what district mathematics specialists and coaches need to know and be able to do to design and facilitate high-quality teacher professional development, and how they can be supported to develop those perspectives, forms of knowledge, and practices.[8] We contend that this was a worthwhile issue to address as the provision of high-quality teacher professional development is an important aspect of district capacity for instructional improvement. The team that conducted this design study in collaboration with the secondary mathematics director was relatively small and the funding required was modest. It is similarly feasible for a small team to conduct many of the studies we have suggested.

As an illustration, in chapter 7, we argued that additional studies are needed that investigate the practices of accomplished coaches, and their influence on teachers' learning. For example, a study that focuses on the practices of a small number of carefully selected coaches might be appropriate and could be conducted by a small team and would require only limited external funding. This team might then conduct follow-up studies that investigate designs for supporting coaches' development of accomplished practices.[9] As a second example, we indicated in chapter 13 that studies are needed that clarify what central office leaders need to know and do in order to support the development of coherent instructional systems at the school level that are organized around rigorous goals for students' learning. Studies that focus on a small number of district leaders would make an important contribution in this regard.

The theory of action that we have reported is relevant to researchers conducting these and other studies, preferably in partnership with practitioners, because it allows them to situate the issues they are addressing within the context of a comprehensive instructional improvement effort. For example, in chapter 7 we shared findings indicating that coaches' relationships with school leaders are critical in enabling them to spend the bulk of their time working with teachers to support their development of ambitious and equitable instructional practices. An investigation of accomplished coaches' practices and of their influence on teachers' learning might therefore select accomplished coaches who work with supportive principals, with the understanding that it would be essential when working in other settings to also consider how to support school leaders' learning. In the case of studies of what central office leaders need to know to support the development of coherent instructional systems, we reported earlier in this chapter and in chapter 13 that principals' perceptions of what members of the Leadership department are holding them accountable for influence the extent to which they implement strategies aimed at supporting teachers' development of ambitious and equitable practices. This suggests that it would be wise in a study of principal supervisors to partner with districts in which members of the leadership and curriculum and instruction departments have compatible agendas that focus on improving the quality of instruction so that students can attain rigorous learning goals.

Investigating and Supporting the Process of Implementation

We clarified at the beginning of chapter 1 that our research goal was to identify a set of interrelated improvement strategies that aim at ambitious and equitable instructional practices. As we have discussed, the resulting theory of action can inform the development of school and district instructional improvement plans. However, as we reported in chapter 14, our partner districts took up 67 percent of the recommendations we made for revising their improvement strategies but struggled to implement the recommendations successfully. The findings reported in the chapters that focused on the various components of the theory of action clarify that the districts were often able to implement the form but not the intended function of improvement strategies. For example, we reported in chapter 6 that while school leaders implemented teacher collaborative groups, the work of most groups was unlikely to support the participating teachers' development of ambitious and equitable instructional practices. In other words, the form, teacher collaborative meetings, was implemented but the intended function, supporting the development of ambitious and equitable practices, was not realized. As a second example, we reported in chapter 12 that principals did, for the most part, make classroom observations and give teachers feedback on their instruction but that the feedback was unlikely to support their development of ambitious and equitable practices. Once again, the form but not the intended function was implemented.

Our partner districts' struggles with implementation are emblematic of the challenges that many other districts are likely to encounter when implementing a range of improvement strategies. The recommendations that our partner districts found difficult to implement successfully were not particularly radical but instead focused on common types of improvement strategies that are being widely implemented. Furthermore, given the criteria we used to identify the four districts, it is likely that their leadership was less accomplished that that of many other districts. These observations indicate that although the identification of potentially productive improvement strategies is an important contribution, it is not sufficient. It is also necessary to figure out what it takes to implement potentially productive strategies reliably in a range of different school and district contexts. In our view, addressing this issue constitutes the next step in a research program that aims to understand what it takes to improve the quality of teaching and student learning on a large scale.

In this regard, we see considerable value in the use of improvement science methods in education as pioneered by the Carnegie Foundation for the Advancement of Teaching under the leadership of Anthony Bryk.[10] Improvement science methods were specifically developed to investigate and improve the implementation of a wide range of initiatives in other fields including medicine. In addition to studies that aim to identify and further clarify potentially productive improvement strategies, we envision inquiries conducted in close collaboration with practitioners that seek to design, analyze, and improve the implementation of specific strategies. For example, the two illustrative cases discussed in chapter 6 clarify the characteristics of effective teacher collaborative meetings. Future investigations might involve researchers collaborating with teachers, coaches, and district math specialists to investigate how to support the participating teachers in working together in these ways. In light of the Carnegie Foundation's strong recommendation to start small and work through the challenges encountered when the focal improvement strategy is implemented on an increasingly wider scale, there is clearly a place for investigations in which small teams of researchers with limited funding work closely with practitioners.

Research on Teaching and Learning, and on Policy and Leadership

The theory of action we have proposed and, more generally the systems perspective that it reflects, spans the traditional concerns of research on classroom teaching and learning, and on educational policy and leadership. It is therefore unfortunate that, on many college campuses, researchers in these two broad areas of inquiry are unfamiliar with each other's work and fail to talk to each other. This schism parallels the divide that we observed between the district central office departments of leadership and of curriculum and instruction (see chapter 13). The schism is consequential in that research on teaching and learning typically treats classrooms as existing in an institutional vacuum despite the abundant evidence that teachers' instructional practices are profoundly influenced by the school and district settings in which they work.[11] As a consequence, research on classroom teaching typically has little to say about supports for instructional improvement that extend beyond issues of curriculum, classroom instruction, and teacher professional development, and research findings are rarely situated within the context of a comprehensive instructional improvement effort. For its part, research in educational policy and leadership frequently treats the classroom as a black box and fails to take a position on worthwhile learning goals and thus on what counts as high-quality instruction. Consequently, the resulting recommendations for school and district improvement strategies are typically generic and often vaguely focused.

The findings we have reported in this book clarify why this schism is highly problematic. Researchers in policy and leadership are right to claim that school and district leadership matters for teaching and learning even if they often fail to specify how it matters. Researchers in teaching and learning are right to claim that student learning goals and thus what counts as high-quality instruction matters for the practices of school and district leaders, even if they often fail to take account of school and district leaders' practices when reporting findings and making recommendations. As we have indicated, it is perfectly acceptable for researchers to focus on a single aspect of an encompassing instructional improvement system. However, in our view, it is also essential that researchers are aware that the element on which they are focusing is part of a system, and have some understanding of how it influences and is influenced by other components of the system. Researchers in teaching and learning and in policy and leadership might then appreciate the relevance of each other's work to their own investigations and might consider collaborating on studies that span the schism between their respective areas of research and expertise.

Relationship Between Research and Practice

As we have indicated, there were major gaps in the research base on which we could build when we began working with our four partner districts in 2007. Although there has been significant progress in some areas, particularly pull-out professional development and instructional coaching, significant gaps remain. For example, there is still little research that can inform school and district leaders' decisions about supplemental supports that enable currently struggling students to participate fully in and learn from mainstream instruction that aims at rigorous learning goals. Similarly, research on what specifically school and district leaders might do to support the development of school and district capacity for instructional improvement remains thin. As researchers, we find it embarrassing that research can currently offer only limited guidance on these and other pressing issues that practitioners routinely have to address in the course of their work.

We contend that this situation is unlikely to improve unless there is a fundamental change in the relationship between research and practice.[12] Currently, issues considered to be worthy of investigation are typically identified exclusively within the tightly bounded echo chamber of particular research communities. When researchers aim to contribute to improvement, they usually work in schools either to test the effectiveness of or to trial and perhaps refine innovations that they have developed as next steps in their research programs independently of the concerns of practitioners. Consequently, there is no "there" there in much educational research because it lacks a grounding in the reality of schooling.

In our view, educational research needs to be at the service of practitioners' instructional improvement efforts if it is to provide much needed empirically grounded guidance. This requires that researchers work with practitioners in ways that prioritize the development of trust, take schools' and districts' current improvement goals and strategies as a primary point of reference, and are sensitive to schools' and districts' capacities and constraints. Researchers who work with practitioners in this manner would strive to see things from their practitioner colleagues' points of view. They might then realize that research has little to offer school and district leaders as they make decisions about supplemental supports for currently struggling students. They might also observe that many districts are investing scarce resources to develop curriculum frameworks and work with practitioners to investigate whether these investments are paying off. In addition, they might become aware that in the absence of research-based guidance, many school and district leaders are struggling in their efforts to build capacity for instructional improvement.

The developments that we have described require a change in dispositions for many researchers such that they begin to not merely understand but to identify with the types of problems that practitioners have to address and the challenges that they encounter in the course of their work. Conversations within research communities might then be broadened and enriched by the voices of practitioners who press to ensure that research agendas are grounded in and are responsive to the reality of schooling.

Afterword

Fort Worth Independent School District: Context And Charge

MICHAEL SORUM

IN 2005, DR. MELODY JOHNSON, the newly hired superintendent of The Fort Worth Independent School District (FWISD) engaged the District in a partnership with the Institute for Learning (IFL), a component of the Learning Research and Development Center (LRDC) at The University of Pittsburgh. When she arrived in Fort Worth, Johnson already knew that the district faced myriad systemic and social challenges that were negatively impacting student achievement and that, despite good will, the district and its staff did not have the systems or capacity to effectively address declining student achievement and a rapidly growing and changing student demographic.

During the decade before Johnson's arrival, the demographics of the district evolved from being predominately Anglo with a stable African American population of about 20 percent to an enrollment that is almost 70 percent Hispanic, 20 percent African American, and 10 percent White with a very small percentage of Asian and "other" student groups. Around 33 percent of the district's 88,000 students are English language learners and around 8 percent of students are identified as special education. Changes in state funding formulas combined with the growth in the district resulted in budget cuts that had virtually eliminated essential support staff and systems such as an up-to-date curriculum, professional development, and the technology infrastructure that would normally support these systems. Every major system in the district, from transportation, to budget, to human resources, to school leadership and curriculum and instruction needed modernization and develop additional capacity.

Johnson had worked with the IFL in her previous district and knew that it had the knowledge, capacity and experience to help modernize teaching and learning in urban districts. The IFL also worked with a national network of experts who were passionate about and engaged in improving outcomes for urban students. It was through the IFL that Dr. Paul Cobb and his MIST team from Vanderbilt University approached Johnson about partnering with the district to support improved outcomes in middle school mathematics, one of the district's lowest performing areas of student achievement.

In 2006, while there were hundreds, if not thousands, of researchers and research organizations working in public school systems across the country, the concept of these researchers and districts engaging in a mutually beneficial partnership was new, or at least not clearly identified and articulated. I knew from having conducted studies in public schools myself as a master's and doctoral student that researchers have good intentions when they engage with school systems. However, my subsequent experience as a senior administrator in two school systems had cooled my receptivity to the idea of "one more researcher" in our district. Upon arriving in the District, I inherited a wide array of research projects but I did not have the sense that these projects were helping me with the problems I faced leading the development of a Division of Teaching and Learning with the goal of improving student outcomes.

DEVELOPING A RESEARCH PRACTICE PARTNERSHIP

My initial reaction to Johnson's suggestion that I speak with Dr. Cobb was quite hesitant. I had a good knowledge of the instructional problems the district faced and I was nearly overwhelmed trying to rebuild a teaching and learning support staff while also leading the writing of a comprehensive curriculum that integrated the state's latest standards. I was at the limits of my "band width" and my visceral reaction to the idea of partnering with MIST researchers was negative. However, using a great deal of tact and patience, Cobb carefully explained his intentions in working with the district and was sufficiently compelling that I opened my mind to the possibility that something good could come from it for the children of the district. A deciding factor for my agreement was that members of the MIST team would do much of the legwork. My staff and I would only need to participate in interviews, provide requested data files, attend partnership meetings, and ensure that the school doors were open to MIST researchers. In retrospect, my decision to engage in the MIST project has become profoundly satisfying to me both professionally and personally. It was exciting to be working at the edge of current researcher practitioner partnership strategies, the children and staff of FWISD benefitted from the work, and I came to welcome the intellectual stimulation that resulted from working with a national network of researchers and practitioners. There have doubtless been effective collaborations between researchers and practitioners in the past, but what did not exist was a framework and the language for explicitly building, engaging in, and replicating successful Research Practice Partnerships (RPPs).

The MIST team elected to work in Fort Worth for at least two significant reasons. The first was a deep confidence in Superintendent Johnson and the board of education to develop and launch an ambitious instructional strategic plan that focused on academic rigor and included both a comprehensive curriculum and ongoing supports for teachers' improvement of their instruction. A second and very important reason was that the district had recently adopted an inquiry-oriented mathematics program, *Connected Mathematics Project 2* (CMP2), which emphasized the development of mathematical understanding, reasoning, and communication as well as procedural fluency.

Working with the IFL, Johnson's team developed instructional leadership protocols and monitoring tools (e.g., Learning Walks) to inform staff about levels of implementation of our strategic plan so that appropriate adjustments could be identified and implemented. To support principals and central administrators in the departments of leadership and of curriculum and Instruction, Johnson's team also developed a new instructional coaching strategy at the middle- and high-school levels called lead content teachers, and hired and developed mathematics and literacy coaches at the elementary level.

The design called for principals to recruit an experienced and successful mathematics teacher at each campus to become the school's lead content teacher (LCT) for mathematics. The LCTs would teach half time and then coach their peers the other half of the school day. The fact that LCTs were actual teachers of the actual students at particular schools was important in creating credibility and accessibility. We hypothesized that middle-school teachers would respond better to changes suggested by colleagues who they viewed as successful internal experts than they would to external experts who might not be familiar with the context of and challenges at each campus. The design also called for school-level administrative support through the principal or an assistant principal monitoring and supporting mathematics instruction. Finally, the district secondary mathematics director led her staff in developing new curriculum frameworks, and leadership directors guided principals in supporting its implementation. The intent of the curriculum framework (CF) was to ensure that all teachers had access to a curriculum that encompassed the written, taught, and tested components of the State's standards. The CF also contained differentiations for ESL, bilingual, special education, and gifted students. The LCT program and curriculum frameworks were launched at around the time that the district began working with MIST.

The value of MIST, and what came to be our RPP, started to become evident after the first full year of implementation and the subsequent "digestion" of the first year's findings. In the first and subsequent years, MIST repeatedly found evidence of a quite clearly voiced coherence of the elements of our strategic plan, LCTs, curriculum frameworks, etc., from all levels of the organization. However, although we had a comprehensive and clearly articulated strategic plan, the findings revealed that we could talk the proverbial walk but were having a lot of trouble walking the talk. There was evidence of significant disconnects between our strategic plan and reality. Longitudinally, MIST data also revealed incremental improvements and successes that would not have been evident if we had focused only on student assessment results. Below, I discuss just a few of the learnings that would have been unlikely without MIST, its protocols, and its data.

LEARNINGS AND OUTCOMES REVEALED THROUGH MIST PARTNERSHIP

Classroom Instruction

Teachers struggled mightily with moving beyond traditional forms of mathematics instruction by developing classroom practices that were consistent with ambitious and equitable mathematics teaching and learning. They specifically struggled with letting the students "get messy" in the learning and eventually, through classroom discussions, prompting, and occasionally correction, developing and deepening their own knowledge and understanding of complex mathematical concepts. While not ill intended, the overwhelming tendency of teachers was to "take" the learning away from students out of a fear either of students' perceived lack of ability or of failure on the part of the teacher. One must also acknowledge the intense pressure that teachers felt to "cover the curriculum" to prepare students for success on state accountability tests. Over time, we did observe improvements in the practices of some teachers, but these improvements were too few and far between.

Perceptions of Students

MIST data revealed what appeared, on the surface, to be an intriguing conundrum. Over the course of several years, middle school achievement in mathematics gradually improved. However, teachers reported decreased levels of comprehension on the part of their students! In our annual meetings to discuss MIST findings, we hypothesized that as teachers began to use more ambitious instructional strategies, they were gaining a more in-depth understanding of what their students did and did not understand. In the past, the teachers "took the learning" from the students and gave them the answers. This resulted in a false sense of achievement, as evidenced by student achievement results. Teachers became more aware of their students' lack of understanding, while in fact, their actual understanding was improving.

Leadership and Learning

As mentioned above, a core tenet of our strategic plan was the importance of school administrators supporting their teachers in improving their instruction. A district expectation was that the principal or an assistant principal attend teacher collaborative meetings and support the teachers during instruction by providing feedback. By and large this happened. However, on campuses where an assistant principal was assigned to the math department, we observed a distinct decline in the teachers' perception of their principal as the instructional leader of their campus. Rather than interpreting this as "positive or negative," we urged principals to take this into consideration and to ponder the associated implications. A result was that on several of the lowest performing campuses at which the principal had delegated responsibility for the math department, the principal changed course and personally assumed the leadership for mathematics.

Another important observation regarding leadership was the role of central office Leadership Directors who worked with principals to support the implementation of *Connected Mathematics Project 2* and the strategies promulgated by the IFL, the district mathematics department, and our MIST partnership. Effectively supporting more ambitious instruction required a significant (higher than in the past) degree of understanding of high-quality mathematics instruction on the part of leadership directors. Without this understanding, or at a minimum, sufficient understanding to let teachers and students try to struggle through the learning, it was very tempting for the administrator to "step in" and try to "save" the situation, which then "took" the learning away from campus administrators and teachers. In one case, a Leadership Director required his schools to implement "scripted" mathematics lessons developed by individuals who did not understand the mathematics required for students to perform well on state assessments. To their great credit, Leadership Directors responded positively to this feedback by becoming more engaged in the MIST work, by attending meetings and debriefings, and also by seeking, more than in the past, input from the mathematics department before making intervention-type actions at the campus level.

Instructional Support

MIST data revealed findings that caused us to significantly change our strategic plan with respect to instructional support for teachers. Our plan rested on the assumption that the LCTs and had a deep understanding of the instructional practices that we wanted to see teachers develop as they implemented CMP2. A sobering finding was that for most LCTs, the depth of their knowledge and sophistication of their instructional practices were not significantly more advanced than those of the teachers they coached. The district employed parallel instructional support strategies for the other three tested content areas and the other LCTs had been selected in a similar manner to the math LCTs. We were therefore concerned that the majority of LCTs, across the subject areas and grade levels, were probably not equipped to successfully coach teachers in improving their instructional practices. This prompted us to take a new direction in selecting and providing professional development for LCTs.

MIST data also revealed the need for significantly more curriculum support and expertise at the district level. But the findings about the content depth of LCTs prompted us to be much more discerning in the selection of curriculum specialists who wrote the curriculum frameworks. This of course created the corresponding challenges caused by "hiring away" effective campus teachers to central office specialist positions. Thus, although we were able to use MIST data successfully to leverage additional mathematics content expertise of district curriculum staff, replacing these highly proficient teachers was problematic.

The selected findings above (there were many, many more) did not reveal unexpected challenges, but they certainly revealed the profundity of the challenges the district faced. The challenges were always there, but the data changed central administrators' perception of their own skills as well as their strategies for selecting and hiring staff who were qualified to do the caliber of work required to implement our strategic plan effectively. The MIST data, in fact, began to really make clear the true depth and complexity of implementing rigorous, meaningful, and lasting systemic instructional improvements. The practices of MIST (observations, interviews, feedback processes) would not allow us to put our heads in the sand and ignore the problems.

Student Achievement During and Since MIST

While MIST supported and informed myriad changes in practices associate with teaching, leading, and learning, the question remains, "Did it improve student outcomes?"

Analyzing eleven years of data, I can say without hesitation that middle school mathematics achievement in FWISD improved over the time the district was involved in the MIST partnership.[1] However, there are numerous strands of data and improvements were not universal. Texas has an extensive data system that allows researchers to ask many questions. Because there were significant changes in state standards and in the format and rigor of the Texas assessment and accountability systems during the partnership, the data that are the most informative are those that compare the rank of performance of students in districts across the state. To determine progress, Fort Worth and the other large urban districts in the state compare their performance against similar districts. Fort Worth uses Austin, Dallas, Houston, and San Antonio as these districts have the most similar demographics.

For the purposes of this discussion, I look specifically at the "exit" scores for eighth grade, the last year of middle school. I will be referring to both TAKS and STAAR scores. The TAKS test, while more challenging than its predecessors, was primarily a test of skills. The STAAR was aligned with revised and much more challenging standards, and included more challenging items designed to assess reasoning and conceptual understanding, and that required students to solve non-routine problems.

As mentioned earlier, Hispanic students are Fort Worth's largest student demographic group, by far. In 2005, and for several years after, at the first administration of the Exit Level (required for promotion) TAKS and STAAR tests, Hispanic students in FWISD consistently ranked at the bottom of these five districts. However, over time and despite more challenging standards, a harder test, and fewer students taking the eighth STAAR because they had been accelerated to Algebra I, the rank of FWISD eighth graders slowly rose. During the last three cycles of STAAR testing, FWISD eighth grade Hispanic students were ranked second and then first for the last two test administrations among our urban peers. Establishing causality in an environment that if influenced by dozens of variables is very difficult. However, in this case, it is a fact that during MIST and in the two years of data since MIST concluded, the ranked performance of eighth grade Hispanic students in FWISD moved from worst to best. By no measure is this insignificant. That said, the relative performance of African American and White students has not changed significantly.

THOUGHTS AND CONSIDERATIONS

Effecting changes in professional practice that result in improvement in student outcomes is not an easy task. There are as many variables to consider, including characteristics of students, teachers, administrators, and school climates. After thirty years as an urban educator, the last twenty of which I was actively involved in and responsible for effecting change, I believe this more than is ever the case. Student outcomes will not improve unless professional practices improve. I challenge anyone who doubts this or doubts the complexity of the task to think of the difficulty of changing their own, or their spouses' or children's behaviors . . . losing ten pounds, walking five times a week, putting the cap on the toothpaste, picking up socks, not to mention practicing the piano or spelling or math. Now, think about supporting and motivating improvements in teachers' and principals' practices, individuals over whom central office administrators have only limited influence. While teachers and school administrators often expressed fears of "getting in trouble," it is uncommon to find a school administrator, or especially a teacher, whose professional position has been damaged by not implementing rigorous instructional tasks.

While it might be non-traditional to discuss relationships within the context of research, the story would not be complete without sharing a couple of brief observations. RPPs are a non-traditional research strategy so it not surprisingly leads to non-traditional discussions and findings. It was personally gratifying for me as a leader to observe my curriculum and leadership staff building positive working relationships with their peers at the district and campus level. All leaders have a "relationship" of some sort with their peers or subordinates but I have often found those relationships to be ones of compliance. Our RPP encouraged meaningful exchanges about deep and personal beliefs associated with teaching and learning. These exchanges were often not pleasant and rarely "easy." However, if one works from the assumption that leadership, instructional, and curricular improvement strategies should be implemented with fidelity, my experiences suggest that fidelity is much more likely to happen in an environment that values communication and trust, and allows for active dialogue about both current beliefs and fears about the future.

Principals and teachers do not truly embrace reforms handed down by a district unless they understand how and why a reform can help them, and they see a way in which they can grow and learn themselves. In this pursuit, a well-designed and implemented RPP is an extremely promising strategy. As a senior leader in a large and complex system, I can truly say that our RPP was a significant variable in the equation that led to me stay in the same position for eleven years. The MIST RPP contributed to the continuity and stability that is essential to sustained and meaningful change.

In my career, I have often found that we have the expectation that "good" data will change outcomes. Outcomes will not change without good data, but good data alone do not guarantee that we will identify and implement the changes in professional practice that are necessary to improve outcomes. While vexing, the alternative of not obtaining or not accepting data is even less likely to improve outcomes. In my opinion this conundrum is the core of the true value of a Research Practice Partnership. A healthy RPP fosters the creation of a learning community that expands beyond the boundaries of the district. External researchers bring a depth and breadth of knowledge that the average educational leader, myself certainly included, simply does not have the time and, in some cases, the interest, to obtain. Melding the extensive practitioner experience of district staff with the knowledge and "science" of researchers creates a space in which true learning can take place; learning that increases the capacity and performance of district and campus leaders, teachers, and most importantly students. The results of traditional research practices often take years to be available, and they are rarely situationally relevant or useful. RPPs allow, at least in research terms, almost immediate and relevant feedback that a district can use to reinforce or revise its improvement strategies. That was certainly the case in Fort Worth as seen in the examples above.

The positive association that the district had with the MIST RPP eventually led to the District engaging in two other major RPP projects: one addressing elementary literacy and the other addressing the scale up and scale out of the practices of successful high schools. Those were both initiatives that were equally valuable and important for the district.

The introduction to this afterword provided some insight into the context that led to FWISD engaging in the MIST RPP. I provided this context because I believe that when organizations are considering engaging with a district in a RPP, they should do so with caution and armed with a solid understanding of the history, context, and dynamics of the district. Launching an RPP is exciting and intellectually stimulating, but it is not an insignificant task; under the best of circumstances, there are very real challenges to the work.

An effective RPP depends on strong relationships but there must be systemic factors that assist in moving the RPP forward. I advise researchers considering a RPP to start by "entering at the top." By this I mean, find a champion and ally at the most senior administrative level. In our case, by the time of its official launch, the MIST RPP had the full support of the Superintendent and the chief academic officer/deputy superintendent. While I would never advocate advancing the work through coercion to compliance, there is a fine line to walk. For mid-level and campus staff, it is helpful for them to know that the project has the unequivocal support of senior staff and even the school board. In our case, a Memorandum of Understanding between Vanderbilt and the District provided a quasi-policy that endorsed both the implementation of CMP2 and the RPP. An RPP can simply not be "forced." It must develop as a support to the district's clearly public strategic plan. It must have clear goals that orient the work while also being iterative and flexible. In an urban school system, there will always be changes and distractions, movements of the foundation, such as changes in board or administrative leadership. These changes will shake even the best partnership. But I firmly believe that a RPP that is built on trust and is aligned with a district's strategic plan will survive, albeit with changes in its structure and form. For extensive research on the components of effective RPPs, in general, I refer those interested in RPPs to the seminal works of William Penuel and Cynthia Coburn referenced in the bibliography.[2]

APPENDIX

How Educational Leaders See MIST: A Case for Long-Term, Mutualistic Partnerships Between Educators and Researchers

ANNA-RUTH ALLEN AND WILLIAM R. PENUEL

THE CHAPTERS IN THIS BOOK have described how the partnerships between researchers and educators in MIST districts developed, as well as their theory of action about how to improve the quality of mathematics instruction at the scale of a large district. Here, we shift the focus to the perspectives of educational leaders in the MIST partner districts. Specifically, we describe how they viewed the partnership with researchers and how they evaluated the work of their partnership for the district. Our aim is to provide a complementary, outsider perspective that synthesizes the common themes while also identifying variations in educational leaders' perspectives on their partnerships with the MIST research team.

Our account draws on data from a separate study, funded by the William T. Grant Foundation, that we conducted of the MIST partnerships with two districts and of one other long-term research practice partnership. The goal of that larger study was to explore how research evidence is developed and used in partnerships where researchers focus their efforts on persistent problems of educational practice facing districts. Studying a partnership like MIST—where researchers were able to work across multiple grants for an extended period of time, and where relationships were built and rebuilt after turnover and organizational change—provides an opportunity to examine what is possible when researchers and educators work together over the long haul on problems of practice.

WHY STUDY RESEARCH-PRACTICE PARTNERSHIPS?

The MIST Project is one of a number of recent efforts underway today to reimagine ways that researchers and practitioners can work together in education. Research-practice partnerships are long-term collaborations between practitioners and researchers that are organized to investigate problems of practice and solutions for improving schools, districts, and other educational organizations.[1] These efforts stem in part from an enduring concern about the usefulness and relevance of research to educational practice.[2] In part because external funding for research studies is largely peer-reviewed—meaning evaluated by other researchers and not by practitioners—researchers have a strong incentive to focus on building academic knowledge and theory. They have fewer incentives to spend time with practitioners, identifying problems of practice that can serve as starting points for developing new lines of research that have both scholarly and practical value.[3] MIST's unique model of long-term engagement with both practical and scholarly aims makes it a valuable project to study to learn more about new forms of collaboration among researchers and educators.

Another inspiration for studying research-practice partnerships as concerted efforts to transform the relationship of research and practice comes from policymakers seeking to promote more "evidence-based practice" in education. Policies like No Child Left Behind and its successor, the Every Student Succeeds Act, emphasize the importance of leaders using evidence from high-quality research when selecting programs and policies.[4] These policies are premised on the idea that improving access to and use of evidence in educational decision making can help leaders make better decisions that ultimately lead to better outcomes for students.[5] A growing body of research supports the idea that sustained interactions with researchers can help educational leaders to make sense of data and research, illuminate problems in new ways, and support the search for solutions to local problems.[6] As a research site for our study, MIST presented an opportunity to examine the role of evidence in a long-term collaboration that focused on problems of practice.

In what follows, we describe our study and present findings about how district leaders characterized the partnership. We conclude with a discussion of some of the key features of these partnerships that we think undergird the themes found in interviews, and play a role in supporting partnership work.

DATA COLLECTION AND METHODS

Our study of research-practice partnerships, conducted with colleagues Cynthia Coburn and Caitlin Farrell, focused on these two research questions: How do researchers and practitioners in partnerships figure out what the focus of their work will be? And how does the work of partnerships shape the use of research evidence for addressing school districts' pressing educational challenges? From 2012 to 2014, we conducted interviews with leaders in the two partner districts, observed both joint meetings with researchers and meetings among district leaders, surveyed district leaders, and analyzed documents produced by the research team and policies developed by district leaders. Here, we narrow the focus to evidence from interviews where we asked leaders about their perceptions of the MIST partnership and the role it played in the district. We explore *how district leaders viewed the roles of the researchers and the consequences of the partnership* as a way to provide a snapshot of what research-practice partnerships can look like.

We conducted interviews twice a year with district leaders and researchers. The interviews explored a range of topics, including perceptions of the partnership, ideas about key problems facing the district, and views about the best strategies the district could pursue to improve the quality of mathematics instruction in all middle school classrooms. We asked district leaders to characterize how they saw MIST researchers and the partnership, including questions like: How would you describe the districts' partnership with researchers on the MIST team? What role do you see them playing in the district? Other interview questions about the details of partnerships work also sometimes prompted reflections on what the research partners brought to the district that leaders found valuable. Over the course of the study, we conducted sixty-seven interviews with district leaders in MIST districts, and of these, 40 were with leaders in District B and twenty-seven were with leaders in District D.

We analyzed interview data from district leaders to develop an understanding of how they saw the MIST researchers and characterized the relationship. We identified all the segments within interviews where leaders explicitly characterized the role played by the MIST researchers, or commented on the consequences or benefits of the partnership for the district. There were eighty-four such segments across interviews with thirty-one district leaders (some interviews did not contain relevant excerpts for this analysis). We then inductively generated categories to use for coding how leaders characterized both the role and the consequences of their research partners. As part of the analysis of roles, we analyzed interview data inductively to identify key *images* of that relationship and the MIST researchers evident in and across the two district partnerships.

FINDINGS

Our analysis showed that there were many ways leaders understood the role of the partnership, and the benefits of their work together. For the leaders we interviewed, the nature and the value of the MIST partnership were multifaceted, and their perceptions were rooted in two overarching qualities of relationship we see evident in the interviews. The first quality is the MIST researchers' concern for the problems that were facing their district. They described researchers as not motivated by self-interest or research questions that are not shared by practitioners, but by a genuine concern for the issues facing their district, and for the policies and programs the district had put in place to address those issues. The second quality is the longevity of the relationship that supported the building of trust. Leaders said that the fact that MIST researchers stuck around helped build relationships gave them reason to trust that researchers were involved to help them improve outcomes for the long haul.

District Leaders' Perceptions of the External Partners

We found there were four primary ways that leaders characterized the MIST partnership (see table A1). Almost half of leaders' descriptions involved characterizations of the partnership as a team of researchers who are *Outside Reflectors* that provide objective research analyses to the district to help

TABLE A1 Perceptions of the external research partners

ROLE	DEFINITION	DISTRICT B	DISTRICT D
Outside Reflector	Provides objective and comprehensive research findings about middle school mathematics in the district	15	12
Thought Partner	Provides opportunities for discussion; guidance about next steps; help with problem-solving, design	13	8
Professional Development Provider	Offers PD within the district for principals and coaches	7	4
Broker	Helps district access to national experts, resources, research	2	2

leaders see what was currently taking place in middle school mathematics classrooms. A third of the leaders characterized the partnership as outside researchers who function as *Thought Partners* and who brought expertise about teaching and learning in mathematics to problem-solving discussions and offered guidance about next steps. A few others described MIST as *Professional Development Providers* who led or co-led sessions for coaches or principals. Finally, a few saw MIST as *Brokers* who helped the district access to national experts, resources, and research. The pattern of findings, as table A1 below shows, is similar across the two districts. The counts in the tables reflect excerpts from interviews coded by category.

Below, we explore these different roles and illustrate them with specific examples of how leaders characterized and perceived the MIST research partners.

Outside reflector

This was the most common characterization of MIST researchers. District leaders talked about the value of hearing perspectives from outside on the districts' own policies, strategies, and instructional practice. They particularly valued the systematic data on classroom practice MIST collected, analyzed, and brought back to district leaders, and several administrators characterized these activities as holding up a "mirror" to central office leaders about current math instructional practice and improvement efforts. One cabinet-level leader from District D asserted,

> One of the things that helps us is to have some outside folks come in and take a completely non-biased look at what's going on . . . Vanderbilt's got no dog in the fight. They're not trying to sell a program. It's kind of like they're able to hold up a mirror and let us see. Because sometimes in the busy-ness of everything, it's hard to look at yourself. I just think it makes the reflection process better.

In this description, the leader characterizes Vanderbilt as "non-biased" because the researchers have "no dog in the fight," that is, that leader sees the researchers as not seeking to get the district to adopt a particular program or practice. Similar descriptions coded in this category call out MIST researchers as "honest" and "coming in and taking an outside look" at districts.

Notably, not all district leaders saw the reflection as neutrally offered empirical data but as one that brought new perspectives to them. One leader in District B called the MIST researchers "a fresh set of eyes" who bring a "different lens" to problems. Another leader from this district said that the researchers brought "a new lens . . . on what we're doing, what they see from their perspective." This view, we think, is in accord with MIST researchers' views of themselves as focused on the district's own initiatives but from a theoretically-informed, empirically-grounded perspective on what it takes to improve the quality of mathematics instruction at the scale of a district. For district leaders who describe MIST researchers as outside reflectors, it was their status as outsiders that they say helped them see things afresh; this may have been because the leaders were too immersed in the work to notice certain aspects of practice or because it is difficult to learn from teachers honestly or systematically about the issues they are facing, because of their authority as district leaders.

Other district administrators who described the MIST researchers as outside reflectors commented on their capacity to play this role this because of their longitudinal approach. One administrator from District B particularly valued the fact that the researchers could track progress across multiple years:

> I really appreciate the fact that we're able to get such a clear view, a user-friendly understanding of what's going on in our classrooms. I think if that's something Vanderbilt can continue to do, and provide, which I

> believe they do, from the very first year, a continuation of how it's going. "Let's take a look at this year compared to all the other years."

To us, this last characterization of the MIST researcher speaks to the possibilities enabled by the long-term nature of the partnership.

Thought partner

The second most common way leaders characterized MIST researchers were as thought partners who engaged in meaning-making, problem-solving, and design work to support mathematics instructional improvement. One leader in District D appreciated the expertise of the researchers but emphasized that they were willing to work with the district in planning policy strategies. He said

> They actually bring "how to," they actually bring the theory and they can bring it into practice. I mean like they can take what they're learning or what they have found, and they can bring it into what's going on [well], what's not working, and what might be good next steps. Then they volunteer to help with the next steps. Help. What else do you want in a partnership?

Other leaders described how the research team served as a "critical friend" and reflected that the partners "don't come and tell us what we need to do, they help us figure out what we need to do." One District B cabinet level leader described how the researchers enriched their conversations and understandings of instructional improvement efforts:

> I can't think of a question, but we make a remark, [and they ask] "Why do you think this happened?" Kind of pushing it a little bit. "Why was that?" It takes you to another level of thinking. You can't just give one answer, you have to figure it through. "Why do you think so?" It's going to another level. "I guess if I really thought about it, this is something else." It's a deeper conversation.

Bringing outside perspectives and analyses of research evidence was part of these leaders' descriptions of the external partnerships, but a willingness to help with the districts' own challenges and policy strategies was also central to characterizations of the MIST team as a thought partner.

Professional development provider

When asked about the MIST team's role, some participants talked about their leadership in professional development with coaches and/or principals. One leader from District D described how the researchers "came and did training with all the staff developers, and then we got to see what they presented to the principals so that everybody was on the same page. The principals knew exactly what to expect and what the focus was and everything. I have gotten a lot out of [the professional development sessions] in the past several years." Some district leaders who foregrounded MIST's role in professional development also described the professional development as co-designed and co-implemented with the district. A District B leader described the professional development activity as one in which "they [MIST researchers] worked with me to develop that and to orient where I wanted to go and where we needed to go with that." One District D leader, for example, described the professional development as co-led "with the developers." At the same time, this leader also commented that the professional development helped "push what we were trying to do," an idea consistent with MIST researchers as critical friends for district leaders. Notably, the leaders who were most likely to highlight this critical friend role were content leaders who participated in co-design work with MIST researchers.

Broker

A less frequently cited characterization we coded as "broker." These descriptions were explicit references to how the research team "broker[ed] services from others, like University of Pittsburgh," especially early in the partnership, or comments about how the researchers were available to provide district leaders with research about particular issues relevant to instructional improvement in mathematics at scale. For example, one math leader in District B said:

> I was talking to [MIST partner]. She's going to look at some of the research and what it says are the key components of common planning time. Being able to say, 'Do you have some research on this? I'm really struggling with the teachers on this. Is there research that can back me up?' Having that sounding board really helps.

Notably, this example is one in which the research asked for and provided focused on something the leader was already working on—and that had not initially been a central focus of the partnership work, but became so over time as use of common planning time became a more important focus within the district and to the research team.

Perceptions of the Value of the Partnership

Embedded in many of the descriptions of partnership activities were explicit evaluations of the consequences or contributions of partnership, that is, statements about what benefit the partnership offered the district (see table A2). The most common benefit across the two districts was in *prompting changes in thinking or learning among district leaders.* A second common benefit reported was *prompting changes in district practice,* though this was more evident in District B than in District D. A third value of the partnership was that it *provided evidence or guidance relevant to deciding next steps (recommendations).* A fourth also relates to evidence but specifically to *evidence that confirms district leaders' existing knowledge.* Finally, a fifth category of value relates to the activity of co-designing and implementing professional development and relates specifically to the *value of the professional development provided.*

As we did in the section above, we illustrate these themes in district leader's accounts of the consequences or benefits of the partnership with examples.

TABLE A2 Perceived benefits of the partnership

ROLE	DEFINITION	DISTRICT B	DISTRICT D
Prompted changes in thinking/learning	New learning, changes in perspectives or knowledge based on partnership work	15	12
Prompted changes in district practice	Partnership work shifted practices of district leaders, principals, coaches, or teachers	11	7
Provided guidance relevant to helping district move forward	Findings and recommendations that helped solidify direction or redirect efforts	9	5
Provided evidence that confirms district leaders' existing knowledge	Vanderbilt's findings and/or input aligned with, confirmed, or validated what the district heard from other partners, or knew from their own experience	7	8
Provided valuable professional development	Benefited from professional development within the district, specifically for principals and coaches	4	6

Prompted changes in thinking/learning

One of the ways that that district leaders said that the partnership benefited them was in expanding their own thinking, particularly about mathematics instruction and how to support improvements in it. One District D leader described coming to appreciate more deeply the importance of mathematical content knowledge among instructional support staff, and commented, "I was like, 'Oh, but wait a second. Actually these coaches, to be more effective in math, they have to have some math background'." Another leader said that the partnership helped them realize "how important the practice standards [of the Common Core State Standards] are" as a guide to instructional planning in mathematics, and that they had not spend enough time working on them with teachers and coaches. In both these cases, leaders attributed their realizations to their participation in the feedback meetings held by Vanderbilt, which were designed opportunities to make sense of the researchers' findings and recommendations.

A content leader in District B who was already familiar with ideas from an earlier engagement with an external partner, the Institute for Learning based at the University of Pittsburgh, commented that she learned a lot from the Vanderbilt team about how to "use those assessing and advancing questions within the classroom so that you can facilitate mathematical discourse." Notably, she characterized the mode of learning not as didactic, but rather as "very, very supportive, collaborative, but not enabling [as in, making them dependent in a negative sense], like allowing us to take it and do it. In other words, they don't do it. We are like a true partner."

Prompted changes in district practice

One of the benefits cited by a number of leaders in both districts was that partnership work led to changes to district practices. Most often, these were related to practices regarding the guidance for instruction, as this leader in District B noted:

> Because they're [strategies emphasized by the MIST team] all good, effective teaching strategies, [we] could incorporate those into science [instruction]. And the other is that I put in place the fact that principals had to be THE administrator overseeing the mathematics work. That way it puts them at the forefront as far as having conversations with teachers, doing classroom visits, sitting in on department meetings, things like that.

Another content leader in District B noted that one example of a shift in district practice that had resulted from the partnership was that senior district leaders committed to engage in professional learning activities, whereas in the past they had only made a "showing" but did not monitor or participate actively. The impetus, this leader said, was the need for coherence of instructional messaging. That leader noted that the "leadership level is really taking an active part in being there for the professional development, making sure that we do have one voice, that we are saying the same thing, that what is being said in one department is going to be echoed by another department." The theme of coherence was echoed in District D, where one leader noted that a key outcome of the partnership was to enhance professional development for principals related to common instructional strategies and to integrate the MIST work with another major initiative in the district.

Leaders attributed these changes in practices to different sources or partnership activities. Some attributed the changes to the reports that the researchers produced, to the fact that they "come in and look around and pull their data, analyze it, and come back with their information." To others, direct observation of professional development activities co-led by the researchers helped to spread particular ideas about good instruction beyond mathematics.

Provided guidance relevant to helping district move forward

One of the ways that leaders say the MIST partnership has benefited their district was by helping them make progress on their own goals in the face of challenges. As a leader in District B said,

> They take our experiences and all the red tape and everything that happens here and work with us to work past those things. We discuss things on what research says, how we can change those things and make sure that we're moving in the direction that we want to be moving in.

Another leader in this district noted that change is a constant within districts, and that the partnership evolved so that certain "pieces" of work or strategies could be tried and reflected on as part of helping the district become a "working district." Some leaders commented on how different the partnership was from other kinds of arrangements between the district and external partners, such as consultants or researchers who approach the district wanting to conduct a study. As one leader in District B put it,

> One way that Vanderbilt differs, usually others who come in have a set plan of how they're going to be incorporated within the structure that's already here. Vanderbilt is more hands-off, "Let's see what you're doing. Let's look at every piece that's going through. Let us come back and give you the analysis of the data that we've gathered. Here are our findings from this. We know that you have your own initiatives. *Here are our suggestions of things that can blend in with them.*"

Another leader in District B commented that part of what the partnership brought to the district were multiple perspectives on problems that the district was facing to inform district changes. In District D, leaders said the partnership helped them figure out what to do to "move instruction" to become more ambitious, so that students would perform better on more challenging state assessments.

Provided evidence that confirms existing knowledge

At least some of what MIST researchers presented at district meetings and in reports confirmed district leaders' assessments of instructional challenges or supported the directions selected leaders proposed their district should move. As a leader in District B said about the district feedback meetings, "For me, I always love these meetings, because I'm a firm believer in that you don't know what you don't know and you can't see what you can't see. Having you-all do this work, it brings validity to the things that we are doing, in some cases." A leader in District D noted that they were already aware of one of the key findings in a report regarding more and less effective ways to work with assessment data in teacher collaborative time. The MIST report "reinforced" some of the directions that were "already in the works," according to this leader. Leaders were quick to point out that although all the reports confirmed what they already believed, some also pushed them in new directions. But in these instances when leaders named areas of congruence with MIST's empirical analyses and recommendations, they treated this convergence as confirming and validating.

A couple of leaders attributed the validating aspects of the recommendations to the structure of the MIST partnerships and the process through which the researchers learned about district priorities. In particular, one

District D leader cited the researchers' ability to facilitate confidential conversations among diverse stakeholders as key to the success:

> Vanderbilt comes in and has all of these face-to-face confidential conversations with all stakeholders, and that they can come back and share it in a report and make recommendations for us that then we can discuss and validate for ourselves, it's not just me talking about the way I think it went, because that's just my point of view. But being able to get that point of view from teachers and assistant principals and principals and content specialists, our leaders, [principal supervisors], getting it from every point of view is very helpful.

Provided valuable professional development

For a handful of leaders, a key benefit of the partnership was the professional development sessions researchers co-planned and co-led for district coaches and principals. One leader in District B described a series of professional development sessions for principals in which the MIST researchers gave principals the task of observing classrooms to look for evidence of cognitive demand of tasks in relation to the lesson structure supported by the curriculum materials. The result was that principals saw for themselves what the researchers' reports had been telling them, that teachers were not making students do enough of the "heavy lifting" in the classroom; instead, "[the teachers] were giving away all the math, or they weren't launching" the lessons in the ways that the curriculum materials prescribed. Leaders also identified professional development sessions and the content of those sessions as a benefit or consequence of the partnership for the district.

Leaders in both districts commented that professional development for principals had helped their districts. A leader in District D noted that most principals followed the "principal as manager" model, but that shifts brought about by Common Core were necessitating the development of principals' content expertise. The professional development that this leader co-designed and co-led with MIST researchers "pointed us in the direction of 'we've got to continue to move forward in this area.'" Similarly, a District B content leader noted that the professional development they had co-designed and co-led was helping the district better prepare its administrators to "become instructional leaders." Finally, a couple of leaders talked about how their participation in co-designing and co-leading the professional development sessions was a valuable part of the partnership for them, particularly because the sessions were embedded in and drew upon other partnership activities that made up the MIST annual data collection, analysis, and feedback cycle.

MAKING SENSE OF HOW LEADERS MAKE SENSE OF THE PARTNERSHIP

Overall, our interview analysis confirmed something we had already begun to sense from observing the MIST team in action and talking informally with district leaders: a wide range of district leaders valued the MIST partnership. Yet they did not just emphasize one role played by the research team or a single benefit. Rather, the roles and value were multifaceted.

There were different ways of characterizing just who the research partners were. To some leaders, the partners were "reflectors" who helped the district see what was going on in an honest way. For some leaders, the external partners were full partners in helping think through the problems associated with improving instruction at the scale of a district. For others less involved in the co-design work, the external partners were viewed as researchers who provided professional development (despite the research partners' varied roles and their intentional efforts to be much more than professional development providers.) Leaders also viewed MIST partners as brokers to ideas and resources from the world of research that were relevant to the problems the district faced. And for some of these leaders, the external partners were viewed as playing more than one of these roles.

Just as with perceptions of roles, leaders named different kinds of benefits or consequences for the district of the partnership. This included both helping confirm leaders' own impressions of what was working in the district and what was not, and also challenging them and helping them see the teaching and learning of mathematics and instructional improvement differently. As one leader put it, the MIST partners did not come in and address an issue that the district knew it had; instead, "They've helped us find our issue." For some, the benefit of the partnership was in the impetus the partnership gave for movement—either in a forward direction that the district was already pursuing, or in a new direction. In this way, the external research partner served as a stimulus and support for leaders to take action. And for still others, the partnership afforded them an opportunity to help design and observe learning opportunities targeted to other leaders in the district. As such, these leaders acquired new models they might apply in future partnership activities or on their own to a core aspect of their work.

Just what accounts for these perceptions? In our analysis, we have highlighted the reasons that leaders themselves gave for their descriptions of the MIST research partners and the benefits of the partnerships. The qualities of the relationship figure prominently in their accounts, as does the annual cycles of collecting and analyzing data to document district practice, reporting findings back to district leaders, and helping them to make sense of the data and its implications for district policy and practice.

Two of the distinguishing characteristics of research-practice partnerships are that they are—*long-term* and *mutualistic.*[7] By *long-term,* we mean that partners commit to work together over a long period of time, typically much longer than a single consulting agreement or grant might last. The long-term nature of partnerships can help build trusting relationships over time, and it can also enable partnerships to tackle big issues in depth.[8] By *mutualistic,* we mean that both researchers and practitioners have a commitment to engage in work that benefits both parties. They negotiate the focus of their work together on an ongoing basis, rather than one party defining the terms of involvement for the other ahead of time as in a more typical arrangement between researchers and practitioners.[9] Mutualism helps ensure that different perspectives—those of practitioners and of researchers—are valued, thereby fostering shared ownership of the work and orienting researchers and practitioners to learn from one another.

Both these qualities are difficult to realize in practice. Grants and consulting agreements tend to be shorter term and limit the commitments that partners are able to make to one another. The trust built up in a partnership can be lost when there is turnover. And it is easy for researchers and educators to talk past each other, making it difficult for them to learn from one another.

Across the district leaders' discussions of MIST's role and the benefits of the partnership, we can see some key qualities that contribute to the principle of mutualism that is essential for research-practice partnerships. For example, one way to characterize the role that MIST researchers played in helping the district reflect on the direction it was heading is that it *guides their seeing* what is taking place in the district. The word "seeing" is key here, and it is a word used by the district leaders that points to their own agency in the process of interpreting what is taking place in the district. But the word "guided" is equally critical for interpreting how it is that district leaders can say that the MIST researchers both confirmed and challenged their existing views. It also reflects that MIST researchers were themselves theoretically driven and drew on an interpretive framework from research to guide their own analyses and recommendations. Thus, they bring part of themselves to the partnership but did not have a pre-set agenda that overshadowed district leaders' concerns, perspectives, and goals. But they did not just make sense of the empirical results alongside leaders, they *give sense* to those findings, grounding them both in their own expertise but also in knowledge of the local district challenges and priorities.[10]

Another aspect of mutualism that we can see from our analysis of interview data is the importance of the flexibility of external partners. Flexibility arises in several different ways, too, such as how researchers respond to and make sense of what the district is trying to do in the context of planning professional development, discussing recommendations, or developing a theory of action. It also arises when the district shifts direction, and the external researchers move with them, from the leaders' perspective. We know through interviews, but also through observations, that the researchers did not stick rigidly to a single plan, but changed tack to a certain extent given the changing needs and dynamics of district leadership. In addition, flexibility arose as a valued quality of the partners in that the researchers were not wedded to a single perspective, but could bring different points of view to the table for consideration. Here again, leaders felt that they had agency with respect to what to do with the information, a signal of the respect MIST researchers had for their sensibilities as leaders.

For leaders who spoke of the annual cycles of collecting and analyzing data, reporting findings, making sense, and planning as key sources of value in the partnership, the quality of partnerships as *intentionally organized* is key. To say that a defining feature of research-practice partnerships is that they are intentionally organized means that they have to develop strategies for supporting mutualism in key activities of partnerships—for negotiating the focus of joint work, uncovering key drivers for improvement, structuring co-design processes, and sharing and interpreting findings from research studies.[11]

The MIST partnership represents a useful case in which to examine the potential of this kind of arrangement for relating research and practice in a way that is different from traditional arrangements between researchers and practitioners. It is important to note that the project operated (and a follow-up project still operates in one of the districts, as of this writing) within the constraints of a funding infrastructure that continues largely to incentivize investigator-initiated, rather than jointly defined, research projects. The partnership remains funded through activities that support specific projects, not open-ended commitments. There may be lessons here for how important it is within that structure to allow teams to have flexibility to change directions over time, to renegotiate the focus of work as priorities shift in organizations.

This account of how district leaders perceived the roles of MIST research partners and the consequence of partnership work for the district has implications for district leaders more generally. Leaders in districts often work with a variety of external partners, including educational consultants, vendors, foundation-based projects, and researchers in higher education. There are likely some overlaps, as well as some very important differences, between the roles and benefits identified by leaders in our study of MIST partnerships, and those of other research teams or even consultants or vendor organizations. The more district leaders know about the varied roles, goals, and partnering arrangements that are possible with external partners, the more strategically they can choose who to work with, in what ways, and towards what ends.

Finally, we think there are implications for the preparation of researchers to engage as external partners in ways that the MIST researchers did. These have to do with both preparing young researchers with a *stance of mutualism,* namely a kind of attitude or openness to being driven by problems of practice and to being flexible in drawing upon a range of theories and findings from across a field, rather than being wedded to one small part of the academic world. It also, though, has to do with providing a toolkit to young researchers for how to organize for mutualism—that is, to develop structures like the annual cycle of inquiry the MIST team assembled and implemented with partner districts—that can ensure that research stays focused on issues of concern to districts while also advancing knowledge and theory. These are tall orders for graduate programs, but we think the MIST case shows the value their approach held for district leaders in our study, and the potential payoff for the investment in change.

Notes

CHAPTER 1

1. National Governors Association Center for Best Practices, Council of Chief State School Officers, *Common Core State Standards* (Washington, DC: National Governors Association Center for Best Practices, 2010); NGSS Lead States, *Next Generation Science Standards: For States, By State*s (Washington, DC: The National Academies Press, 2013).
2. Lorrie A. Shepard, "Classroom Assessment," in *Educational Measurement*, ed. Robert L. Bennan (Washington, DC: National Council on Measurement in Education and American Council on Education/Praeger, 2006), 623–646.
3. William H. Schmidt and Nathan A. Burroughs, "How the Common Core Boosts Quality and Equality," *Educational Leadership* 70, no. 4 (2012): 54–58; William H. Schmidt and Richard T. Houang, "Curricular Coherence and the Common Core State Standards for Mathematics," *Educational Researcher* 41, no. 8 (2012): 294–308.
4. Lauren Dotsen and Virginia Foley, "Middle Grades Student Achievement and Poverty Levels: Implications for Teacher Preparation," *Journal of Learning in Higher Education* 12, no. 2 (2016): 33–44.
5. James W. Stigler and James Hiebert, *The Teaching Gap* (New York: Free Press, 1999/2009).
6. James I. Hiebert and Douglas A. Grouws, "The Effects of Classroom Mathematics Teaching on Students' Learning," in *Second Handbook of Research on Mathematics Teaching and Learning*, ed. Frank K. Lester, Jr. (Greenwich, CT: Information Age Publishing, 2007), Vol. 1, 271-405.
7. Deborah L. Ball and David K. Cohen, "Developing Practice, Developing Practitioners: Toward a Practice-based Theory of Professional Education," in *Teaching as the Learning Profession: Handbook of Policy and Practice*, eds. Linda Darling-Hammond and Gary Sykes (San Francisco: Jossey Bass, 1999), 3–32; Ruth Chung Wei et al., *Professional Learning in the Learning Profession: A Status Report on Teacher Development in the United States and Abroad* (Dallas, TX: National Staff Development Council, 2009).
8. This account of the NUMMI joint venture is based primarily on an episode of the National Public Radio show *This American Life*. A transcript of the episode is available at: www.thisamericanlife.org/radio-archives/episode/561/transcript. Wikipedia also proved useful: en.wikipedia.org/wiki/NUMMI.
9. Megan Loef Franke et al., "Capturing Teachers' Generative Change: A Follow-up Study of Teachers' Professional Development in Mathematics," *American Educational*

Research Journal 38 (2001): 653–689; Paola Sztajn, Hilda Borko and Thomas Smith, "Research on Mathematics Professional Development," in *Compendium for Research in Mathematics Education*, ed. Jinfa Cai (Reston, VA: National Council of Teachers of Mathematics, 2017), 793–823.
10. Chris Argyris and Donald A. Schön, *Theory in Practice: Increasing Professional Effectiveness* (San Francisco: Jossey-Bass, 1974); Chris Argyris and Donald A. Schön, *Organizational Learning: A Theory of Action Perspective* (Reading, MA: Addison Wesley, 1978).
11. We draw directly on the notion of a coherent instructional system as proposed by Fred Newmann and colleagues, and Anthony Bryk and colleagues. Our findings indicate how this general notion should be customized when a school or district aims to support all students' attainment of rigorous learning goals. Fred M. Newmann et al., "Instructional Program Coherence: What It Is and Why It Should Guide School Improvement Policy," *Educational Evaluation and Policy Analysis* 23 (2001): 297–321; Anthony S. Bryk et al., *Organizing Schools for Improvement: Lessons From Chicago* (Chicago: University of Chicago Press, 2010).
12. Personal communication, October, 2011.
13. Richard Elmore discusses the problems that arise when improvement is equated with change. See Richard F. Elmore, "Leadership as the Practice of Improvement," (paper presented at the *OECD International Conference on Perspectives on Leadership for Systemic Improvement*, London, England, July 6, 2006).
14. Larry Cuban's seminal historical analysis documents that ongoing changes in policies intended to impact teaching is very much the norm. Larry Cuban, *How Teachers Taught: Constancy and Change in American Classrooms* (New York: Longman, 1984). Frederick M. Hesse coined the term "policy churn" to describe this phenomenon. Frederick M. Hesse, *Spinning Wheels: The Politics of Urban School Reform* (Washington, DC: The Brookings Institute, 1999).

CHAPTER 2

1. Sean Larsen et al., "Understanding the Concepts of Calculus: Frameworks and Roadmaps Emerging from Educational Research," in *The Compendium for Research in Mathematics Education*, ed. Jinfa Cai (Reston, VA: National Council of Teachers of Mathematics, 2017), 526-550; Joanne Lobato and C. David Walters, "A Taxonomy of Approaches to Learning Trajectories and Progressions," in *The Compendium for Research in Mathematics Education,* ed. Jinfa Cai (Reston, VA: National Council of Teachers of Mathematics, 2017), 74-101; Ana C. Stephens et al., "A Learning Progression for Elementary Students' Functional Thinking," *Mathematical Thinking and Learning, 19* (2017): 143-166.
2. Gwendolyn M. Lloyd, Jinfa Cai, and James E. Tarr, "Issues in Curriculum Studies: Evidence-based Insights and Future Directions," in *The Compendium for Research in Mathematics Education,* ed. Jinfa Cai (Reston, VA: National Council of Teachers of Mathematics, 2017), 824–852; James E. Tarr et al., "The Impact of Middle-Grades Mathematics Curricula and the Classroom Learning Environment on Student Achievement, "*Journal for Research in Mathematics Education* 39, no. 3 (2008): 247–280.
3. Megan Loef Franke, Elham Kazemi, and Daniel Battey, "Mathematics Teaching and Classroom Practice," in *Second Handbook of Research on Mathematics Teaching and Learning,* edited by Frank K. Lester (Greenwich, CT: Information Age Publishers,

2007), 225–256; Kara Jackson et al., "Exploring Relationships Between Setting Up Complex Tasks and Opportunities to Learn in Concluding Whole-Class Discussions in Middle-Grades Mathematics Instruction," *Journal for Research in Mathematics Education* 44, no. 4 (2013): 646–682; Victoria R. Jacobs and Denise A. Spangler, "Research on Core Practices in K–12 Mathematics Teaching," in *The Compendium for Research in Mathematics Education*, (Reston, VA: National Council of Teachers of Mathematics, 2017), 766–792; Mary Kay Stein et al., "Orchestrating Productive Mathematical Discussions: Five Practices for Helping Teachers Move Beyond Show and Tell," *Mathematical Thinking and Learning* 10, no. 4 (2008): 313–340.

4. Hilda Borko et al., *Mathematics Professional Development: Improving Teaching Using the Problem-Solving Cycle and Leadership Preparation Models* (New York: Teachers College Press, 2015); Magdalene Lampert et al., "Using Designed Instructional Activities to Enable Novices to Manage Ambitious Mathematics Teaching," in *Instructional Explanations in the Disciplines*, eds. Mary Kay Stein and Linda Kucan (New York: Springer, 2010), 129–141; Paola Sztajn, Hilda Borko, and Thomas Smith, "Research on Mathematics Professional Development," in *Compendium for Research in Mathematics Education*, ed. Jinfa Cai (Reston, VA: National Council of Teachers of Mathematics, 2017), 793–823.
5. James Stigler and James I. Hiebert, *The Teaching Gap: Best Ideas From the World's Teachers for Improving Education in the Classroom* (New York: Free Press, 1999).
6. For notable exceptions, see Barbara S. Nelson and Annette Sassi, *The Effective Principal: Instructional Leadership for High-quality Learning* (New York: Teachers College Press, 2005); Mary Kay Stein and James P. Spillane, "Research on Teaching and Research on Educational Administration: Building a Bridge," in *A New Agenda for Research on Educational Leadership,* eds. William A. Firestone and Carolyn Riehl (Thousand Oaks, CA: Sage Publications, 2005), 28-45.
7. For notable exceptions, see Cynthia E. Coburn and Jennifer Lin Russell, "District Policy and Teachers' Social Networks," *Educational Evaluation and Policy Analysis* 30, no. 3 (2008): 203-235; David K. Cohen, *Teaching and Its Predicaments* (Cambridge, MA: Harvard University Press, 2011); James P. Spillane and Charles L.Thompson, "Reconstructing Conceptions of Local Capacity: The Local Education Agency's Capacity for Ambitious Instructional Reform," *Educational Evaluation and Policy Analysis,* 19 (1997): 185–203.
8. See also Meredith I. Honig, "District Central Office Leadership as Teaching: How Central Office Administrators Support Principals' Development as Instructional Leaders," *Educational Administration Quarterly*, 48 (2012): 733–774.
9. For a discussion of the defining characteristics of RPPs and of the key distinctions between different types of RPPs, see Cynthia E. Coburn, William Penuel, and Kimberly E. Geil, *Research-practice Partnerships: Strategies for Leveraging Research for Educational Improvement in School Districts* (New York: William T. Grant Foundation, 2013).
10. RPPs of this type are called *Research Alliances* and typically focus on a specific school district or a single region. For example, the Consortium on Chicago School Research is a longstanding alliance between the University of Chicago, Chicago Public Schools, and other local community organizations (https://consortium.uchicago.edu). As a second example, Regional Educational Laboratories funded by the Institute of Education Sciences are typically composed of multiple research

alliances that aim to support schools and districts in their geographic regions in using data and research to improve academic outcomes for students.

11. The five central office departments are curriculum and instruction, leadership, special education, English language learning, and research and evaluation. MIST is an example of a Design Research Partnership. Most partnerships of this type aim to support teachers' development of specific instructional practices that have been linked empirically to student learning in a particular content area.
12. For discussions for the key characteristics of productive partnerships between researchers and practitioners, see Cynthia E. Coburn and William Penuel, "Research-practice Partnerships in Education: Outcomes, Dynamics, and Open Questions," *Educational Researcher* 45 (2016) 45–54. 2016; William Penuel, Cynthia E. Coburn, and Daniel J. Gallagher, "Negotiating Problems of Practice in Research-practice Design Partnerships," in *Design-based Implementation Research: Theories, Methods, and Exemplars. One Hundred and Twelfth Yearbook of the National Society for the Study of Education,* ed. Barry J. Fishman, et al. (Chicago: National Society for the Study of Education, 2013), 237–255.
13. We are grateful to the Institute for Learning at the University of Pittsburgh for introducing us to the districts and for facilitating their recruitment.
14. Glenda Lappan et al., *Connected Mathematics Project 2* (Boston: Pearson, 2009).
15. The number of schools in each district depended on the size of middle-grades schools and thus the number of mathematics teachers in each school.
16. Richard F. Elmore, *School Reform from the Inside Out* (Cambridge, MA: Harvard Education Press, 2004).
17. David K. Cohen and Carol A. Barnes, "Pedagogy and Policy," in *Teaching for Understanding: Challenges for Policy and Practice*, eds. David K. Cohen, Milbrey W. McLaughlin and Joan E. Talbert (San Francisco: Jossey Bass, 1993), 207–239.
18. Coburn and Russell, "District Policy"; Cohen and Barnes, "Pedagogy"; Lea Hubbard, Hugh Mehan and Mary Kay Stein, *Reform as Learning* (New York: Routledge, 2008).
19. Jean Lave and Etienne Wenger, *Situated Learning: Legitimate Peripheral Participation* (New York: Cambridge University Press, 1991); Barbara Rogoff, "Evaluating Development in the Process of Participation: Theory, Methods, and Practice Building on Each Other," in *Change and Development: Issues of Theory, Application, and Method*, eds. Eric Amsel and Ann Renninger (Hillsdale, NJ: Erlbaum, 1997), 265–285; Anna Sfard, *Thinking as Communicating: Human Development, the Growth of Discourses, and Mathematizing* (New York: Cambridge University Press, 2008).
20. Anthony S. Bryk, "Support a Science of Performance Improvement," *Phi Delta Kappan* 90 (2009) 597–600; Spillane and Thompson, "Reconstructing."
21. John S. Brown and Paul Duguid, "Organizational Learning and Communities-of-practice: Towards a Unified View of Working, Learning, and Innovation," *Organizational Science* 2 (1991): 40-57; Barbara Rogoff, "Developing Understanding of the Idea of Communities of Learners," *Mind, Culture, and Activity* 1 (1994): 209–229.
22. Jean Lave, "The Practice of Learning," in *Understanding Practice,* eds. Seth Chaiklin and Jean Lave (Cambridge: Cambridge University Press, 1993), 3–32.
23. Martha S. Feldman and Brian T. Pentland, "Reconceptualizing Organizational Routines as a Source of Flexibility and Change," *Administrative Science Quarterly* 48 (2003), 94–118.

24. Magdalene Lampert and Filippo Graziani, "Instructional Activities as a Tool for Teachers' and Teacher Educators' Learning," *Elementary School Journal* 109 (2009): 491–509; James P. Spillane and Jennifer Z. Sherer, "Constancy and Change in Work Practice in Schools: The Role of Organizational Routines," *Teachers College Record* 113 (2011): 611–657.
25. David K. Cohen and Heather C. Hill, "Instructional Policy and Classroom Performance: The Mathematics Reform in California," *Teachers College Record* 102 (2000): 294–343; Janine T. Remillard, "Examining Key Concepts in Research on Teachers' Use of Mathematics Curricula," *Review of Educational Research* 75 (2005): 211–246; James P. Spillane, "External Reform Initiatives and Teachers' Efforts to Reconstruct Their Practice: The Mediating Role of Teachers' Zones of Enactment," *Curriculum Studies* 31 (1999): 143–175.
26. A detailed description of the process we developed for analyzing transcripts of the two hundred interviews in a relatively short period of time can be found in Erin Henrick, Paul Cobb, and Kara Jackson, "Educational Design Research to Support System-wide Instructional Improvement," in *Approaches to Qualitative Research in Mathematics Education: Examples of Methodology and Methods,* eds. Angelika Bikner-Ahsbahs, Christine Knipping, and Norma Presmeg (New York: Springer, 2015), 497–530.
27. Melissa D. Boston, "Assessing the Quality of Mathematics Instruction," *Elementary School Journal* 113, no. 1 (2012): 76–104; Lindsay C. Matsumura et al., "Toward Measuring Instructional Interactions 'at-scale'," *Educational Assessment* 13 (2008): 267–300.
28. Heather C. Hill, Stephen G. Schilling, and Deborah L. Ball, "Developing Measures of Teachers' Mathematics Knowledge for Teaching," *The Elementary School Journal* 105, no. 1 (2004): 11–30.
29. Deborah L. Ball, Mark H. Thames, and Geoffrey C. Phelps, "Content Knowledge for Teaching: What Makes it Special?" *Journal of Teacher Education* 59, no. 5 (2008): 389–407.
30. Chris Argyris and Donald A. Schön, *Theory in Practice: Increasing Professional Effectiveness* (San Francisco: Jossey-Bass, 1974); Chris Argyris and Donald A. Schön, *Organizational Learning: A Theory of Action Perspective* (Reading, MA: Addison Wesley, 1978).
31. A detailed discussion and illustration of the process of formulating improvement recommendations can be found in Paul Cobb and Kara Jackson, "Analyzing Educational Policies: A Learning Design Perspective," *The Journal of the Learning Sciences* 21, no. 4 (2012): 487–521.
32. A discussion of our initial conjectures can be found in Paul Cobb and Thomas Smith, "District Development as a Means of Improving Mathematics Teaching and Learning at Scale," in *International Handbook of Mathematics Teacher Education: Vol. 3. Participants in Mathematics Teacher Education: Individuals, Teams, Communities and Networks,* eds. Konrad Krainer and Terry Wood (Rotterdam, The Netherlands: Sense, 2008), 231–254.
33. Coburn and Russell, "District Policy."
34. Paul Cobb et al., "Design Experiments in Education Research," *Educational Researcher* 32, no. 1 (2003): 9–13; Design-Based Research Collaborative, "Design-based Research: An Emerging Paradigm for Educational Inquiry," *Educational Researcher* 32 no. 1 (2003), 5–8.

35. Paul Cobb, Kara Jackson, and Charlotte Dunlap Sharpe, "Conducting Design Studies to Investigate and Support Mathematics Students' and Teachers' Learning," in *Compendium for Research in Mathematics Education,* ed. Jinfa Cai (Reston, VA: National Council of Teachers of Mathematics, 2017), 208–236.
36. For a discussion of the adaptations we made to the design research methodology to use it at the district level, see Paul Cobb, Kara Jackson, and Charlotte Dunlap, "Design Research: An Analysis and Critique," in *Handbook of International Research in Mathematics Education,* eds. Lyn D. English and David Kirshner, 3rd ed. (London: Taylor & Francis, 2016), 481–503; Paul Cobb, Kara Jackson, Thomas Smith, Michael Sorum, and Erin Henrick, "Design Research with Educational Systems: Investigating and Supporting Improvements in the Quality of Mathematics Teaching and Learning at Scale," in *Design-based Implementation Research: Theories, Methods, and Exemplars. One Hundred and Twelfth Yearbook of the National Society for the Study of Education,* ed. Barry J. Fishman, et al. (Chicago: National Society for the Study of Education, 2013), 320–349.
37. See Coburn et al., *Research Practice Partnerships* for a discussion of different types of research-practice partnerships. The kind of research conducted by such partnerships has sometimes been called design-based implementation research (DBIR): William R. Penuel et al., "Organizing Research and Development at the Intersection of Learning, Implementation, and Design," *Educational Researcher,* 40, no. 7 (2011): 331–337. DBIR involves "a focus on persistent problems of practice" from the perspectives of both practitioners and researchers, and "a commitment to iterative, collaborative design" (p. 332).
38. For a description of the *Instructional Quality Assessment Mathematics Toolkit*, see Melissa D. Boston, "Assessing the Quality." For a description of the *Mathematical Knowledge for Teaching* assessment, see Heather C. Hill et al., "Developing Measures."For a description of the assessment of teachers' visions of high-quality mathematics instruction, see Charles Munter, "Developing Visions of High-Quality Mathematics Instruction," *Journal for Research in Mathematics Education* 45, no. 5 (2014): 584–635. For a description of the assessment of teachers' views of their students' mathematical capabilities, see Kara Jackson, Lynsey Gibbons, and Charlotte Sharpe, "Teachers' Views of Students' Mathematical Capabilities: Challenges and Possibilities for Ambitious Reform," *Teachers College Record* 118, no. 8 (2017).
39. Glenda Lappan, James T. Fey, William M. Fitzgerald, Susan N. Friel, and Elizabeth Difanis Phillips, *Connected Mathematics Project 2* (Boston: Pearson, 2009).
40. See Boston, "Assessing the Quality" for more information about each of the rubrics.
41. Jackson et al., "Exploring Relationships."
42. Deborah L. Ball et al., "Content Knowledge for Teaching."
43. Hill et al., "Developing Measures."
44. Heather C. Hill, "Mathematical Knowledge of Middle School Teachers: Implications for the No Child Left Behind Policy Initiative," *Educational Evaluation and Policy Analysis* 29, no. 2 (2007): 95–114.
45. Paul Cobb, Kay McClain, Teruni Lamberg, and Crystal Dean, "Situating Teachers' Instructional Practices in the Institutional Setting of the School and District," *Educational Researcher* 32, no. 6 (2003): 13–24.

46. Cobb and Smith, "District Development."
47. Munter, "Developing Visions."
48. James P. Spillane, "Cognition and Policy Implementation: District Policymakers and the Reform of Mathematics Education," *Cognition and Instruction* 18, no. 2 (2000): 141–179.
49. Ibid.
50. For examples of studies that suggest that teachers' instruction depends, in part, on their views of their students' capabilities, see: John B. Diamond, Antonia Randolph, and James P. Spillane, "Teachers' Expectations and Sense of Responsibility for Student Learning: The Importance of Race, Class, and Organizational Habitus," *Anthropology and Education Quarterly* 35, no. 1 (2004): 75–98; Ilana S. Horn, "Fast Kids, Slow Kids, Lazy Kids: Classification of Students and Conceptions of Subject Matter in Math Teachers' Conversations," *The Journal of the Learning Sciences* 16 (2007): 37–79; Kara Jackson, "The Social Construction of Youth and Mathematics: The Case of a Fifth Grade Classroom," in *Mathematics Teaching, Learning, and Liberation in the Lives of Black Children*, edited by Danny Bernard Martin, 175–199. New York: Routledge, 2009; Paola Sztajn, "Adapting Reform Ideas in Different Mathematics Classrooms: Beliefs Beyond Mathematics," *Journal of Mathematics Teacher Education* 6 (2003): 53–75.
51. For example, see Diamond et al., "Teachers' Expectations"; Jackson, "The Social Construction"; Sztajn, "Adapting Reform Ideas."
52. For a full description of the assessment, see Jackson et al., "Teachers' Views."
53. National Council of Teachers of Mathematics, *Principles to Actions: Ensuring Mathematical Success for All*. Reston, VA: National Council of Teachers of Mathematics, 2014.
54. Jackson et al., "Exploring Relationships."
55. Jo Boaler and Megan Staples, "Creating Mathematical Futures Through an Equitable Teaching Approach: The Case of Railside School," *Teachers College Record* 110, no. 3 (2008): 608–645; Megan Staples, "Supporting Whole-Class Collaborative Inquiry in a Secondary Mathematics Classroom," *Cognition and Instruction* 25, nos. 2-3 (2007): 161–217.

CHAPTER 3

1. We would like to thank, in alphabetical order, Melissa Boston, Jason Brasel, Paul Cobb, Rip Correnti, Lynsey Gibbons, Emily Kern, Mahtab Nazemi, Brooks Rosenquist, Emily Shahan, Charlotte Sharpe, Thomas Smith, and Jonee Wilson for their contributions to the lines of inquiry reported on in this chapter.
2. National Governors Association Center for Best Practices, Council of Chief State School Officers, *Common Core State Standards* (Washington, DC: National Governors Association Center for Best Practices, 2010); National Council of Teachers of Mathematics, *Principles and Standards for School Mathematics* (Reston, VA: Author, 2000).
3. National Council, *Principles and Standards*; National Council of Teachers of Mathematics, *Principles to Actions: Ensuring Mathematical Success for All*. (Reston, VA: Author, 2014).
4. Kara Jackson and Jonee Wilson, "Supporting African American Students' Learning of Mathematics: A Problem of Practice," *Urban Education* 47, no. 2 (2012): 354–398.
5. From our perspective, equity is not inherent to any form of instruction. However,

particular forms of practice may make it more likely that more students will be supported to participate substantially in rigorous mathematical activity. Thus, the goal of this line of inquiry was to identify forms of practice that had the *potential* to support broader, substantial participation.

6. In describing this vision of teaching as ambitious, we follow other mathematics education researchers, for example, Magdalene Lampert, Heather Beasley, Hala Ghousseini, Elham Kazemi, and Megan Loef Franke, "Using Designed Instructional Activities to Enable Novices to Manage Ambitious Mathematics Teaching," in *Instructional Explanations in the Disciplines*, eds. Mary Kay Stein and Linda Kucan (New York: Springer, 2010), 129–141.
7. Mary Kay Stein, Barbara W. Grover, and Marjorie Henningsen, "Building Student Capacity for Mathematical Thinking and Reasoning: An Analysis of Mathematical Tasks Used in Reform Classrooms," *American Educational Research Journal* 33, no. 2 (1996): 455–88.
8. Mary Kay Stein and Suzanne Lane, "Instructional Tasks and the Development of Student Capacity to Think and Reason: An Analysis of the Relationship Between Teaching and Learning in a Reform Mathematics Project," *Educational Research and Evaluation* 2, no. 1 (1996): 50–80.
9. Jo Boaler and Megan Staples, "Creating Mathematical Futures Through an Equitable Teaching Approach: The Case of Railside School," *Teachers College Record* 110, no. 3 (2008): 608–645; Stein and Lane, "Instructional Tasks."
10. Stein and Lane, "Instructional Tasks."
11. Anne Garrison Wilhelm, "The Enactment of Cognitively Demanding Mathematical Tasks: Investigating Links to Teachers' Knowledge and Beliefs," *Journal for Research in Mathematics Education* 45, no. 5 (2014): 636–674.
12. Kara Jackson et al., "Exploring Relationships Between Setting Up Complex Tasks and Opportunities to Learn in Concluding Whole-Class Discussions in Middle-Grades Mathematics Instruction," *Journal for Research in Mathematics Education* 44, no. 4 (2013): 646–682; Kara Jackson et al., "Launching Complex Tasks," *Mathematics Teaching in the Middle School* 18, no. 1 (2012): 24–29.
13. Research that has focused on the importance of ensuring that contextual features of a task are accessible to all students includes Deborah L. Ball, Imani Masters Goffney, and Hyman Bass,"The Role of Mathematics Instruction in Building a Socially Just and Diverse Democracy," *The Mathematics Educator* 15, no. 1 (2005): 2–6; Jo Boaler, "Learning From Teaching: Exploring the Relationship Between Reform Curriculum and Equity," *Journal for Research in Mathematics Education* 33, no. 4 (2002): 239–258; and William F. Tate, "Returning to the Root: A Culturally Relevant Approach to Mathematics Pedagogy," *Theory into Practice* 34, no. 3 (1995): 166–173.
14. Megan Loef Franke, Elham Kazemi, and Daniel Battey, "Mathematics Teaching and Classroom Practice," in *Second Handbook of Research on Mathematics Teaching and Learning*, edited by Frank K. Lester (Greenwich, CT: Information Age Publishers, 2007), 225–256; Lampert et al., "Using Designed Instructional Activities."
15. Mary Kay Stein et al., "Orchestrating Productive Mathematical Discussions: Five Practices for Helping Teachers Move Beyond Show and Tell," *Mathematical Thinking and Learning* 10, no. 4 (2008): 313–340.

16. Mary Kay Stein and Margaret S. Smith, *Five Practices for Orchestrating Productive Mathematics Discussions* (Reston, VA: National Council of Teachers of Mathematics, 2011), 20.
17. Erna Yackel and Paul Cobb, "Sociomathematical Norms, Argumentation, and Autonomy in Mathematics," *Journal for Research in Mathematics Education* 27, no. 4 (1996): 458–477.
18. Boaler and Staples, "Creating Mathematical Futures."
19. Paul Cobb, "Theorizing About Mathematical Conversations and Learning From Practice," *For the Learning of Mathematics* 18, no. 1 (1998): 46–48.
20. Elham Kazemi and Deborah Stipek, "Promoting Conceptual Thinking in Four Upper-Elementary Mathematics Classrooms," *Elementary School Journal* 102, no. 1 (2001): 59–80; Stein et al., "Orchestrating Productive Mathematical Discussions."
21. Deborah L. Ball, "Teaching, With Respect to Mathematics and Students," in *Beyond Classical Pedagogy: Teaching Elementary School Mathematics*, eds. Terry Wood, Barbara Scott Nelson and Janet E. Warfield (Mahwah, NJ: Erlbaum, 2001), 11–22.
22. Peter C. Murrell, "In Search of Responsive Teaching for African American Males: An Investigation of Students' Experiences of Middle School Mathematics Curriculum," *Journal of Negro Education* 63, no. 4 (1994): 556–569.
23. Jonee Wilson, Mahtab Nazemi, and Kara Jackson, "Investigating Teaching Practice in Conceptually Oriented Mathematics Classrooms Characterized by African American Student Success," *Journal for Research in Mathematics Education* (under review).
24. For other studies that suggest the importance of positioning students as competent, see, for example Daniel Battey et al., "The Interconnectedness of Relational and Content Dimensions of Quality Instruction: Supportive Teacher-Student Relationships in Urban Elementary Mathematics," *The Journal of Mathematical Behavior* 42 (2016): 1–19; and Franke et al., "Mathematics Teaching and Classroom Practice."
25. James Stigler and James I. Hiebert, *The Teaching Gap: Best Ideas From the World's Teachers for Improving Education in the Classroom* (New York: Free Press, 1999).
26. For a description of the *Instructional Quality Assessment Mathematics Toolkit*, see Melissa D. Boston, "Assessing the Quality of Mathematics Instruction," *Elementary School Journal* 113, no. 1 (2012): 76–104. For a description of the *Mathematical Knowledge for Teaching* assessment, see Heather C. Hill, Stephen G. Schilling, and Deborah L. Ball, "Developing Measures of Teachers' Mathematics Knowledge for Teaching," *The Elementary School Journal* 105, no. 1 (2004): 11–30. For a description of the assessment of teachers' visions of high-quality mathematics instruction, see Charles Munter, "Developing Visions of High-Quality Mathematics Instruction," *Journal for Research in Mathematics Education* 45, no. 5 (2014): 584–635. For a description of the assessment of teachers' views of their students' mathematical capabilities, see Kara Jackson, Lynsey Gibbons, and Charlotte Sharpe, "Teachers' Views of Students' Mathematical Capabilities: Challenges and Possibilities for Ambitious Reform," *Teachers College Record* 119, no. 7 (2017).
27. Boston and Wilhelm, "Middle School Mathematics Instruction."
28. Jackson et al., "Exploring Relationships."
29. Ibid.
30. Boston and Wilhelm, "Middle School Mathematics Instruction."

31. For different analyses within the MIST study we examined individual rubrics, sub-scales, and/or overall IQA scores. Sub-scales were created based on factor analysis of the rubrics which suggested conceptually dividing the rubrics into two groups: cognitive demand and discussion. Overall IQA scores were created by averaging the cognitive demand and discussion sub-scales.
32. Deborah L. Ball, Mark H. Thames, and Geoffrey C. Phelps, "Content Knowledge for Teaching: What Makes it Special?" *Journal of Teacher Education* 59, no. 5 (2008): 389–407.
33. Hill et al., "Developing Measures."
34. For discussion of the MKT for the MIST teacher sample, see Wilhelm, "The Enactment of Cognitively Demanding Mathematical Tasks." Note that in our analyses, we use a combined average of two subtest item response theory scale scores to form a single MKT score. The scale scores are standardized and based on analysis of a nationally representative sample; see Hill, "Mathematical Knowledge." Across our sample, the mean MKT was -0.09. Our sample standard deviation was 0.79, indicating that our data was slightly less spread out than the national norm. However, the variation between districts was significant; the highest district mean was 0.44 and the lowest overall district mean was -0.32, representing approximately three quarters of a standard deviation difference between them.
35. Anne Garrison, "Understanding Teacher and Contextual Factors That Influence the Enactment of Cognitively Demanding Mathematics Tasks" (PhD diss., Vanderbilt University, 2013).
36. Paul Cobb et al., "Situating Teachers' Instructional Practices in the Institutional Setting of the School and District," *Educational Researcher* 32, no. 6 (2003): 13–24.
37. Charles Munter and Richard Correnti, "Examining Relations Between Mathematics Teachers' Instructional Vision, Knowledge, and Change in Practice," *American Journal of Education* 123, no. 2 (2017: 171–202), doi.org/10.1086/689928.
38. Munter, "Developing Visions."
39. For example, as described in Munter, "Developing Visions," the average role of the teacher score for the 44 participants whose interviews were assessed with that rubric every year began at 1.98 (out of 4) in year one and ended at 2.66 in year four—a change of approximately one standard deviation. Likewise, average scores on the patterns/structure of classroom talk rubric grew from 2.18 to 3.03 (out of 4)—just over one standard deviation. A change from level 2 to 3 in describing patterns/structure of classroom talk represents a shift from descriptions of classroom talk being limited to small group settings to suggesting that there should also be whole-class discussion following some mathematical work in which students have just engaged.
40. For examples of studies that suggest that teachers' instruction depends, in part, on their views of their students' capabilities, see: John B. Diamond, Antonia Randolph and James P. Spillane, "Teachers' Expectations and Sense of Responsibility for Student Learning: The Importance of Race, Class, and Organizational Habitus," *Anthropology and Education Quarterly* 35, no. 1 (2004): 75–98; Ilana S. Horn, "Fast Kids, Slow Kids, Lazy Kids: Classification of Students and Conceptions of Subject Matter in Math Teachers' Conversations," *The Journal of the Learning Sciences* 16, no. 1 (2007): 37–79; Kara Jackson, "The Social Construction of Youth and Mathematics: The Case of a Fifth Grade Classroom," in *Mathematics Teaching, Learning, and Liberation in*

the Lives of Black Children, ed. Danny Bernard Martin (New York: Routledge, 2009), 175–199.; Paola Sztajn, "Adapting Reform Ideas in Different Mathematics Classrooms: Beliefs Beyond Mathematics," *Journal of Mathematics Teacher Education* 6, no. 1 (2003): 53-75.

41. For example, see Diamond et al., "Teachers' Expectations"; Jackson, "The Social Construction"; Sztajn, "Adapting Reform Ideas."
42. Jackson et al., "Teachers' Views."
43. Ibid.
44. Ibid.
45. Ibid.
46. Jürgen Baumert et al., "Teachers' Mathematical Knowledge, Cognitive Activation in the Classroom, and Student Progress," *American Educational Research Journal* 47, no. 1 (2010): 133-180; Charalambos Y.Charalambous, "Mathematical Knowledge for Teaching and Task Unfolding: An Exploratory Study," *Elementary School Journal* 110, no. 3 (2010): 247–278, doi: 10.1086/648978; Yasemin Copur-Gencturk, "The Effects of Changes in Mathematical Knowledge on Teaching: A Longitudinal Study of Teachers' Mathematical Knowledge and Instruction," *Journal for Research in Mathematics Education* 46, no. 3 (2015): 280–330.
47. Wilhelm, "The Enactment of Cognitively Demanding Mathematical Tasks."
48. For example, see Alan H. Schoenfeld, *How We Think: A Theory of Goal-Oriented Decision Making and Its Educational Applications* (New York: Routledge, 2011); Laurie Sleep and Samuel L. Eskelson, "MKT and Curriculum Are Only Part of the Story: Insights From a Lesson on Fractions," *Journal of Curriculum Studies* 44, no. 4 (2012): 537–558; and Wilhelm, "The Enactment of Cognitively Demanding Mathematical Tasks."
49. Wilhelm, "The Enactment of Cognitively Demanding Mathematical Tasks."
50. For example, see Deborah L. Ball, "Research on Teaching Mathematics: Making Subject Matter Knowledge Part of the Equation," in *Advances in Research on Teaching: Vol. 2. Teachers' Subject Matter Knowledge and Classroom Instruction*, ed. Jere E. Brophy (Greenwich, CT: JAI Press, 1991), 1–48.; Schoenfeld, *How We Think*; and Sleep and Eskelson, "MKT and Curriculum."
51. Charles Munter, "Envisioning the Role of the Mathematics Teacher," *NCSM Journal of Mathematics Education Leadership* 16, no. 1 (2015): 29–40; Wilhelm, "The Enactment of Cognitively Demanding Mathematical Tasks."
52. Anne Garrison Wilhelm, Charles Munter, and Kara Jackson, "Examining Relations Between Teachers' Diagnoses of Sources of Students' Difficulty in Mathematics and Students' Opportunities to Learn," *Elementary School Journal* 117, no. 3 (2017): 345–370.
53. Munter, "Developing Visions."
54. Munter and Correnti, "Examining Relations."
55. Wilhelm, "The Enactment of Cognitively Demanding Mathematical Tasks."
56. Wilhelm et al., "Examining Relations."
57. Charlotte Dunlap, "Examining How School Settings Support Teachers' Improvement of Their Classroom Instruction" (PhD diss., Vanderbilt University, 2016).
58. For example, see Hill et al., "Developing Measures."
59. Munter and Correnti, "Examining Relations."

60. For research on mathematics teacher noticing, see, for example Miriam G. Sherin, "Developing a Professional Vision of Classroom Events," in *Beyond Classical Pedagogy: Teaching Elementary School Mathematics*, eds. Terry Wood, Barbara Scott Nelson, and Janet Warfield (Hillsdale, NJ: Erlbaum, 2001), 75–93; and Elizabeth A. van Es and Miriam G. Sherin, "Mathematics Teachers' 'Learning to Notice' in the Context of a Video Club," *Teaching and Teacher Education* 24, no. 2 (2008): 244–276.
61. Dunlap, "Examining How School Settings."
62. Lynsey Gibbons and Paul Cobb, "Content-Focused Coaching: Five Key Practices," *The Elementary School Journal* 117, no. 2 (2016): 237–260.
63. Jackson et al., "Teachers' Views."
64. Anthony S. Bryk et al., *Learning to Improve: How America's Schools Can Get Better at Getting Better* (Cambridge, MA: Harvard Education Press, 2015); David Yeager et al., *Practical Measurement* (Stanford, CA: Carnegie Foundation for the Advancement of Teaching, 2013).
65. Kara Jackson et al.,"Practical Measures to Improve the Quality of Small-Group and Whole-Class Discussion," (working paper, 2016). http://www.education.uw.edu/pmr2/files/2016/10/White-paper-practical-measures-160914.docx; see also http://www.education.uw.edu/pmr2/.
66. For a review of the field's understanding of how proportional reasoning develops, see, for example, Susan J. Lamon, "Rational Numbers and Proportional Reasoning: Toward a Theoretical Framework for Research," in *Second Handbook of Research on Mathematics Teaching and Learning*, ed. Frank K. Lester (Greenwich, CT: Information Age Publishing, 2007), 629–667.
67. Megan Loef Franke et al., "Capturing Teachers' Generative Change: A Follow-Up Study of Professional Development in Mathematics," *American Educational Research Journal* 38, no. 3 (2001): 653–689.
68. Daniel Battey and Megan Loef Franke, "Integrating Professional Development on Mathematics and Equity: Countering Deficit Views of Students of Color," *Education and Urban Society* 47, no. 4 (2015): 443–462.
69. Jackson et al., "Exploring Relationships"; Wilson et al., "Investigating Teaching."
70. Daniel Reinholz and Niral Shah, "Equity Analytics: A Methodological Approach for Quantifying Participation Patterns in Mathematics Classroom Discourse," *Journal for Research in Mathematics Education* 49, no. 2 (2018): 140–177.
71. Ibid, 149.

CHAPTER 4

1. Walter Doyle, "Work in Mathematics Classes: The Context of Students' Thinking During Instruction," *Educational Psychologist* 23, no. 2 (1988): 167–180; James W. Stigler and James I. Hiebert, *The Teaching Gap: Best Ideas from the World's Teachers for Improving Education in the Classroom* (New York: Free Press, 1999).
2. Melissa D. Boston and Anne G. Wilhelm, "Middle School Mathematics Instruction in Instrumentally-focused Urban Districts," *Urban Education* 52, no. 7 (2017), 829-861, doi:10.1177/0042085915574528.
3. Paul Cobb and Kara Jackson, "Analyzing Educational Policies: A Learning Design Perspective," *Journal of Learning Sciences* 21, no. 4 (2012): 487–521; David K. Cohen, "A Revolution in One Classroom: The Case of Mrs. Oublier," *Educational Evaluation*

and Policy Analysis 12, no. 3 (1990): 311–329; David K. Cohen and Heather C. Hill, "Instructional Policy and Classroom Performance: The Mathematics Reform in California," *Teachers College Record* 102, no. 2 (2000): 294–343.

4. Ruth Chung Wei et al., *Professional Learning in the Learning Profession: A Status Report on Teacher Development in the United States and Abroad* (Dallas, TX: National Staff Development Council, 2009).
5. Ibid.
6. Dan C. Lortie, *Schoolteacher: A Sociological Study* (Chicago: University of Chicago Press, 1975); Judith Warren Little, "The Persistence of Privacy: Autonomy and Initiative in Teachers' Professional Relations," *Teachers College Record* 91, no. 4 (1990): 509–536.
7. Deborah Loewenberg Ball and David K. Cohen, "Developing Practice, Developing Practitioners: Toward a Practice-based Theory of Professional Education," in *Teaching as the Learning Profession: Handbook of Policy and Practice*, eds. Linda Darling-Hammond and Gary Sykes (San Francisco: Jossey Bass, 1999), 3–32.
8. Deborah Loewenberg Ball et al., "Combining the Development of Practice and the Practice of Development in Teacher Education," *Elementary School Journal* 109, no. 5 (2009): 458–474.
9. Mary Kay Stein and Suzanne Lane, "Instructional Tasks and the Development of Student Capacity to Think and Reason: An Analysis of the Relationship Between Teaching and Learning in a Reform Mathematics Project," *Educational Research and Evaluation* 2, no. 1 (1996): 50–80.
10. Mary Kay Stein et al., "Orchestrating Productive Mathematical Discussions: Five Practices for Helping Teachers Move Beyond Show and Tell," *Mathematical Thinking and Learning* 10, no. 4 (2008): 313–340.
11. Deborah Loewenberg Ball, "Teaching, with Respect to Mathematical and Students," in *Beyond Classical Pedagogy: Teaching Elementary School Mathematics*, eds. Terry Wood, Barbara Scott Nelson and Janet Warfield (Mahwah, NJ: Erlbaum, 2001), 11–22.
12. Stein et al., "Orchestrating Productive."
13. Pam Grossman et al., "Teaching Practice: A Cross-Professional Perspective," *Teachers College Record* 111, no. 9 (2009): 2055–2100.
14. Hilda Borko et al., "Using Video Representations of Teaching in Practice-based Professional Development Programs," *ZDM Mathematics Education* 43, no. 1 (2011): 175–187; Miriam G. Sherin and Sandra Y. Han, "Teacher Learning in the Context of a Video Club," *Teaching and Teacher Education* 20, no. 2 (2004): 163–183; Elham Kazemi and Megan L. Franke, "Teacher Learning in Mathematics: Using Student Work to Promote Collective Inquiry," *Journal of Mathematics Teacher Education* 7, no. 3 (2004): 203–235.
15. Grossman et al., "Teaching Practice."
16. Morva McDonald, Elham Kazemi, and Sarah S. Kavanagh, "Core Practices and Pedagogies of Teacher Education: A Call for a Common Language and Collective Activity," *Journal of Teacher Education* 64, no. 5 (2013): 378–386.
17. Ellice A. Forman, "A Sociocultural Approach to Mathematics Reform: Speaking, Inscribing and Doing Mathematics Within Communities of Practice," in *A Research Companion to Principles and Standards for School Mathematics*, eds. Jeremy Kirkpatrick, W. Gary Martin, and Deborah Schifter (Reston, VA: National Council of Teachers

of Mathematics, 2003), 333–352; Jean Lave and Etienne Wenger, *Situated Learning: Legitimate Perophesal Participation* (London: Cambridge University Press, 1991).

18. Lynsey Gibbons and Paul Cobb, "Content-Focused Coaching: Five Key Practices," *Elementary School Journal* 117, no. 2 (2016): 237–260.
19. Ibid.
20. Britnie D. Kane, "Facilitating Sensemaking in Teacher Workgroups: Concept Development Through Representations of Practice" (in preparation); Susan B. Nolen, Ilana S. Horn, and Chistopher J. Ward, "Situational Motivation," *Educational Psychologist* 50, no. 3 (2015): 234–247; Megan Webster, "A Decomposition of the Practices of High Quality Professional Development Facilitation for Teachers," (PhD diss., McGill University, 2015).
21. Nicole A. Bannister, "Reframing Practice: Teacher Learning Through Interactions in a Collaborative Group," *Journal of the Learning Sciences* 24, no. 3 (2015): 347–372, doi:10.1080/10508406.2014.999196; Rogers Hall and Ilana S. Horn, "Talk and Conceptual Change at Work: Analogy and Epistemic Stance in a Comparative Analysis of Statistical Consulting and Teacher Workgroups," *Mind, Culture and Activity* 19, no. 3 (2012): 240–258; Ilana S. Horn, "Teaching Replays, Teaching Rehearsals, and Re-visions of Practice: Learning from Colleagues in a Mathematics Teacher Community," *Teachers College Record* 112, no. 1 (2010): 225–259.
22. Kara Jackson and Paul Cobb, "Coordinating Professional Development Across Contexts and Role Groups," in *Teacher Education and Pedagogy: Theory, Policy and Practice*, ed. Michael Evans (Cambridge, UK:Cambridge University Press, 2013), 80–99.

CHAPTER 5

1. We would like to thank Paul Cobb, Charlotte Sharpe, Lynsey Gibbons, and Mollie Appelgate, as well as our district leader partners, for their substantial contributions to the line of inquiry focused on pull-out professional development in the MIST project.
2. Deborah L. Ball and David K. Cohen, "Developing Practice, Developing Practitioners: Toward a Practice-Based Theory of Professional Education," in *Teaching as the Learning Profession: Handbook of Policy and Practice*, eds. Linda Darling-Hammond and Gary Sykes (San Francisco: Jossey Bass, 1999), 3–32; Hilda Borko, "Professional Development and Teacher Learning: Mapping the Terrain," *Educational Researcher* 33, no. 8 (2004): 3–15; Paola Sztajn, Hilda Borko, and Thomas Smith, "Research on Mathematics Professional Development," in *Compendium for Research in Mathematics Education*, ed. Jinfa Cai (Reston, VA: National Council of Teachers of Mathematics, 2017), 793–823.
3. Ruth Chung Wei et al., *Professional Learning in the Learning Profession: A Status Report on Teacher Development in the United States and Abroad* (Dallas, TX: National Staff Development Council, 2009).
4. Ball and Cohen, "Developing Practice."
5. Megan Webster, "A Decomposition of the Practices of High Quality Professional Development Facilitation for Teachers," (PhD diss., McGill University, 2016).
6. Ball and Cohen, "Developing Practice."
7. Ibid; Deborah Ball et al., "Combining the Development of Practice and the Practice of Development in Teacher Education," *Elementary School Journal* 109, no. 5 (2009): 458–474.

8. Ball and Cohen, "Developing Practice," 12.
9. Mary Kay Stein, Janine T. Remillard, and Margaret S. Smith, "How Curriculum Influences Student Learning," in *Second Handbook of Research on Mathematics Teaching and Learning*, ed. Frank K. Lester (Greenwich, CT: Information Age Publishing, 2007), 319–371.
10. Pam Grossman et al., "Teaching Practice: A Cross-Professional Perspective," *Teachers College Record* 111, no. 9 (2009): 2055–2100.
11. Ibid.
12. Hilda Borko, Karen Koellner, and Jennifer Jacobs, "Examining Novice Teacher Leaders' Facilitation of Mathematics Professional Development," *Journal of Mathematical Behavior* 33 (2014):149–167, doi: 10.1016/j.jmathb.2013.11.003; Sztajn, Borko and Smith, "Research on Mathematics Professional Development."
13. Lynsey Gibbons and Paul Cobb, "Content-Focused Coaching: Five Key Practices" *Elementary School Journal* 117, no. 2 (2016): 237–260; Ilana S. Horn and Britnie Delinger Kane, "Opportunities for Professional Learning in Mathematics Teacher Workgroup Conversations: Relationships to Instructional Expertise," *Journal of the Learning Sciences* 24, no. 3 (2015): 373–418; Kara Jackson et al., "Investigating the Development of Mathematics Leaders' Capacity to Support Teachers' Learning on a Large Scale," *ZDM Mathematics Education* 47, no. 1 (2015): 93–104, doi: 10.1007/s11858-014-0652-5; Webster, "A Decomposition"; Jonee Wilson, "Investigating and Improving Designs for Supporting Professional Development Facilitators' Learning," (PhD diss., Vanderbilt University, 2015).
14. Eric R. Banilower et al. *Report of the 2012 National Survey of Science and Mathematics Education* (Chapel Hill, NC: Horizon Research, Inc., 2013).
15. For a case of a MIST district in which the quality of professional development was exceptional, see Webster, "A Decomposition."
16. Ball and Cohen, "Developing Practice"; Borko, "Professional Development and Teacher Learning."
17. See Webster, "A Decomposition," for a case of a MIST district in which pull-out professional development was led by a district mathematics leader, and the facilitation was of exceptional quality.
18. Jackson et al., "Investigating the Development"; Wilson, "Investigating and Improving Designs."
19. Wilson, "Investigating and Improving Designs."
20. Jackson et al., "Investigating the Development"; Wilson, "Investigating and Improving Designs."
21. Ball and Cohen, "Developing Practice," 19.
22. Borko, "Professional Development and Teacher Learning"; Katja Maaß and Michiel Doorman, "A Model for Widespread Implementation of Inquiry-Based Learning," *ZDM Mathematics Education* 45, no. 6 (2013): 887–899, doi: 10.1007/s11858-013-0505-7; Sztajn, Borko and Smith, "Research on Mathematics Professional Development."
23. Borko, Koellner, and Jacobs, "Examining Novice Teacher Leaders"; Rebekah Elliot et al., "Conceptualizing the Work of Leading Mathematical Tasks in Professional Development," *Journal of Teacher Education* 60 (2009): 364–379, doi: 10.1177/0022487109341150.
24. Daniel Battey and Megan Loef Franke, "Integrating Professional Development on

Mathematics and Equity: Countering Deficit Views of Students of Color," *Education and Urban Society* (2013): 1–30, doi: 10.1177/0013124513497788.

25. Charlotte Dunlap, "Examining How School Settings Support Teachers' Improvement of Their Classroom Instruction" (PhD diss., Vanderbilt University, 2016); Kara Jackson, Lynsey Gibbons and Charlotte Sharpe, "Teachers' Views of Students' Mathematical Capabilities: Challenges and Possibilities for Ambitious Reform,"*Teachers College Record* 119, no. 7 (2017).
26. Jackson, Gibbons and Sharpe, "Teachers' Views."
27. See, for example, Battey and Franke, "Integrating Professional Development"; Melissa Gresalfi and Paul Cobb, "Negotiating Identities for Mathematics Teaching in the Context of Professional Development," *Journal for Research in Mathematics Education* 42, no. 3 (2011): 270–304.
28. Jackson, Gibbons and Sharpe, "Teachers' Views."
29. Mary Kay Stein and Suzanne Lane, "Instructional Tasks and the Development of Student Capacity to Think and Reason: An Analysis of the Relationship Between Teaching and Learning in a Reform Mathematics Project," *Educational Research and Evaluation* 2, no. 1 (1996): 50–80.
30. Judit Moschkovich, "Supporting the Participation of English Language Learners in Mathematical Discussion," *For the Learning of Mathematics* 19, no. 1 (1999):11-19.
31. See, for example, Jo Boaler and Megan Staples, "Creating Mathematical Futures Through an Equitable Teaching Approach: The Case of Railside School," *Teachers College Record* 110, no. 3 (2008): 608–645; Ilana S. Horn, *Strength in Numbers: Collaborative Learning in Secondary Mathematics* (Reston, VA: National Council of Teachers of Mathematics, 2012).
32. Moschkovich, "Supporting the Participation".
33. Hilda Borko et al., "Video as a Tool for Fostering Productive Discussions inMathematics Professional Development," *Teaching and Teacher Education* 24, no. 2 (2008): 417–436; Borko, Koellner, and Jacobs, "Examining Novice Teacher Leaders"; Elliot et al.,"Conceptualizing the Work."
34. Heather C. Hill et al., "Mathematical Knowledge for Teaching and the Mathematical Quality of Instruction: An Exploratory Study," *Cognition and Instruction* 26, no.4 (2008): 430–511; Charles Munter and Richard Correnti, "Examining Relations Between Mathematics Teachers' Instructional Vision, Knowledge, and Change in Practice," *American Journal of Education* 123, no. 2 (2017: 171–202), doi.org/10.1086/689928; Anne Garrison Wilhelm, Charles Munter, and Kara Jackson, "Examining Relations Between Teachers' Diagnoses of Sources of Students' Difficulty in Mathematics and Students' Opportunities to Learn," *Elementary School Journal* 117, no. 3 (2017): 345–370.
35. Borko, Koellner, and Jacobs, "Examining Novice Teacher Leaders."
36. Ibid, 165; See also Elliot et al., "Conceptualizing the Work"; and Webster, "A Decomposition."
37. Elliot et al., "Conceptualizing the Work"; Pam Grossman, Karen Hammerness and Morva McDonald, "Redefining Teaching, Re-imagining Teacher Education," *Teachers and Teaching: Theory and Practice* 15, no. 2 (2009): 273–289; Jackson et al., "Investigating the Development."
38. Gibbons and Cobb, "Content-Focused Coaching."

39. Ibid; Jackson et al., "Investigating the Development"; Karen Koellner, Jennifer Jacobs and Hilda Borko, "Mathematics Professional Development: Critical Features for Developing Leadership Skills and Building Teacher Capacity," *Mathematics Teacher Education and Development* 13, no. 1 (2011): 115–136.
40. Horn and Kane, "Opportunities for Professional Learning"; Webster, "A Decomposition."
41. Webster, "A Decomposition."
42. Nicole Bannister, "Reframing Practice: Teacher Learning Through Interactions in a Collaborative Group," *Journal of the Learning Sciences* 24, no. 3 (2015): 347–372, doi: 10.1080/10508406.2014.999196; Horn and Kane, *Opportunities for Professional Learning.*
43. Sztajn, Borko, and Smith, "Research on Mathematics Professional Development," 41.
44. Webster, "A Decomposition."
45. Jackson et al., "Investigating the Development"; Wilson, "Investigating and Improving Designs."
46. Hilda Borko et al.,. *Mathematics Professional Development: Improving Teaching Using the Problem-Solving Cycle and Leadership Preparation Models* (New York: Teachers College Press, 2015); Elliot et al., "Conceptualizing the Work."
47. Jackson et al., "Investigating the Development."
48. Wilson, "Investigating and Improving Designs."
49. Ibid.
50. For exceptions, see, for example, Battey and Franke, "Integrating Professional Development"; and Gresalfi and Cobb, "Negotiating Identities."
51. Dunlap, "Examining How School Settings."
52. Hilda Borko, Karen Koellner, and Jennifer Jacobs, "Meeting the Challenges of Scale: The Importance of Preparing Professional Development Leaders," *Teachers College Record,* Date Published: March 04, 2011 http://www.tcrecord.org *ID Number: 16358*; Maaß and Doorman, "A Model"; Sztajn, Borko and Smith, "Research on Mathematics Professional Development."

CHAPTER 6

1. Yvonne L. Goddard, Roger D. Goddard and Megan Tschannen-Moran, "A Theoretical and Empirical Investigation of Teacher Collaboration for School Improvement and Student Achievement in Public Elementary Schools," *Teachers College Record* 109, no. 4 (2007): 877–896; Judith A. Langer, "Excellence in English in Middle and High School: How Teachers' Professional Lives Support Student Achievement," *American Educational Research Journal* 37, no. 2 (2000): 397–439; Milbrey W. McLaughlin and Joan E. Talbert, *Professional Communities and the Work of High School Teaching* (Chicago: University of Chicago Press, 2001); Matthew Ronfeldt et al., "Teacher Collaboration in Instructional Teams and Student Achievement," *American Educational Research Journal* 52, no. 3 (2015): 475–514.
2. Richard DuFour, "What is a Professional Learning Community?" *Educational Leadership* 61, no. 8 (2004): 6–11; Willis D. Hawley and Linda Valli, "The Essentials of Effective Professional Development: A New Consensus," in *Teaching as the Learning Profession: Handbook of Policy and Practice,* eds. Linda Darling-Hammond and Gary Sykes (San Francisco: Jossey-Bass, 1999), 127–150.

3. Thomas H. Levine, "Tools for the Study and Design of Collaborative Teacher Learning: The Affordances of Different Conceptions of Teacher Community and Activity Theory," *Teacher Education Quarterly* 37, no. 1 (2010): 109–130.
4. Robert C. Bogdan and Sari K. Biklen, *Qualitative Research for Education: An Introduction to Theory and Methods*, 3rd ed. (Boston: Allyn and Bacon, 1992).
5. Nicole A. Bannister, "Reframing Practice: Teacher Learning Through Interactions in a Collaborative Group," *Journal of the Learning Sciences* 24, no. 3 (2015): 347–372; Cynthia Coburn, "Making Sense of Reading: Logics of Reading in the Institutional Environment and the Classroom," (PhD diss., Stanford University, 2001).
6. In Chapter Seven, we suggest that this expert facilitator be an instructional coach with expertise in ambitious and equitable mathematics instruction.
7. Betty Achinstein, "Conflict Amid Community: The Micropolitics of Teacher Collaboration," *Teachers College Record* 104 (2002): 421–455; Ilana S. Horn, "Teaching Replays, Teaching Rehearsals, and Re-visions of Practice: Learning from Colleagues in a Mathematics Teacher Community," *Teachers College Record* 112, no. 1 (2010): 225-259; Vicki Vescio, Dorene Ross and Alyson Adams, "A Review of Research on the Impact of Professional Learning Communities on Teaching Practice and Student Learning," *Teaching and Teacher Education* 24 (2008): 80–91.
8. Achinstein, "Conflict Amid Community"; Ilana S. Horn and Judith Warren Little, "Attending to Problems of Practice: Routines and Resources for Professional Learning in Teachers' Workplace Interactions," *American Educational Research Journal* 47, no. 1 (2010): 181–217.
9. Kara Jackson et al., "Exploring Relationships between Setting up Complex Tasks and Opportunities to Learn in Concluding Whole-class Discussions in Middle-grades Mathematics Instruction," *Journal for Research in Mathematics Education* 44, no. 4 (2013): 646–682.
10. Linda Darling-Hammond and Milbrey W. McLaughlin, "Policies That Support Professional Development in an Era of Reform," *Phi Delta Kappan* 76, no. 8 (1995): 597–604; Elham Kazemi and Amanda Hubbard, "New Directions for the Design and Study of Professional Development: Attending to the Coevolution of Teachers' Participation across Contexts," *Journal of Teacher Education* 59, no. 5 (2008): 428–441.
11. Joan Talbert and Milbrey McLaughlin, "Professional Communities and the Artisan Model of Teaching," *Teachers and Teaching* 8, no. 3 (2002): 325–343; Horn, "Teaching Replays."
12. Horn and Little, "Attending to Problems."
13. Brette Garner and Ilana S. Horn, "Using Standardized Test Data as a Starting Point for Inquiry: A Case for Thoughtful Compliance," in *Teachers' Data Use: Cases of Promising Practice*, eds. Nicole Barnes and Helenrose Fives (New York: Routledge, in press).
14. Bannister, "Reframing Practice"; Kara Jackson, Lynsey Gibbons and Charlotte Sharpe, "Teachers' Views of Students' Mathematical Capabilities: Challenges and Possibilities for Ambitious Reform, *Teachers College Record* 118, no. 8 (2017).
15. Horn and Little, "Attending to Problems."
16. Ilana S. Horn, "Fast Kids, Slow Kids, Lazy Kids: Framing the Mismatch Problem in Math Teachers' Conversations," *Journal of the Learning Sciences* 16, no. 1 (2007): 37–79; Judith Warren Little, "The Persistence of Privacy: Autonomy and Initiative in Teachers' Professional Relations," *Teachers College Record* 91, no. 4 (1990):

509–536; Dan Lortie, *Schoolteacher* (Chicago: University of Chicago Press, 1974); McLaughlin and Talbert, *Professional Communities.*

17. To put their reluctance in context, teachers typically do not have enough time to do the urgent daily work of planning and grading, let alone taking on the similarly difficult tasks of rehearsing practice or analyzing students' thinking.
18. Cheryl J. Craig, "Research in the Midst of Organized School Reform: Versions of Teacher Community in Tension," *American Educational Research Journal* 46, no. 2 (2009): 598–619; Tamara H. Nelson et al., "Leading Deep Conversations in Collaborative Inquiry Groups," *The Clearing House* 83, no. 5 (2010): 175–179; R. Bruce Williams, "Four Dimensions of the School Change Facilitator," *Journal of Staff Development* 17, no. 1 (1996): 48–50.
19. Jerome Cranston, "Relational Trust: The Glue that Binds a Professional Learning Community," *Alberta Journal of Educational Research* 57, no. 1 (2011): 59–72; Louise Stoll et al., "Professional Learning Communities: A Review of the Literature," *Journal of Educational Change* 7, no. 4 (2006): 221–258.
20. Jessica G. Rigby, Christine Andrews-Larson and I-Chien Chen, "Teachers' Learning Opportunities about Teaching Mathematics: A Longitudinal Case Study of Administrators' Influence," under review.
21. Christine Andrews-Larson, Jonee Wilson and Adrian Larbi-Cherif, "Instructional Improvement and Teachers' Collaborative Conversations: The Role of Focus and Facilitation," *Teacher College Record* 119, no. 2 (2017): 1–37.
22. Andrews-Larson et al., "Instructional Improvement."
23. Viviane M. J. Robinson and Helen S. Timperley, "The Leadership of the Improvement of Teaching and Learning: Lessons from Initiatives with Positive Outcomes for Students," *Australian Journal of Education* 51, no. 3 (2007): 247–262.
24. Ann Lieberman and Maureen Grolnick. "Networks and Reform in American Education," *Teachers College Record* 98, no. 1 (1996): 7–45.
25. Nelson et al., "Leading Deep Conversations"; Craig, "Research in the Midst"; Williams, "Four Dimensions of the School."
26. In chapter 7, we suggest that co-planning is a potentially productive coaching activity. However, effective co-planning goes beyond discussions of logistics and pacing that we have described in this chapter. Indeed, a key differentiation between effective co-planning and less effective versions is the extent to which teachers and coaches discuss the *why* of instruction.
27. Na'ilah Nasir et al., *Mathematics for Equity: A Framework for Successful Practice* (New York: Teachers College Press, 2014).
28. Pam Grossman and Morva McDonald, "Back to the Future: Directions for Research in Teaching and Teacher Education," *American Educational Research Journal* 45, no. 1 (2008): 184–205.
29. For a study examining the emergence of a teacher collaborative group, see Bannister, "Reframing Practice."

CHAPTER 7

1. Ruth Chung Wei et al., *Professional Learning in the Learning Profession: A Status Report on Teacher Development in the United States and Abroad* (Dallas, TX: National Development Council, 2009).

2. Patricia F. Campbell and Nathanial N. Malkus, "The Impact of Elementary School Mathematics Coaches on Student Achievement," *The Elementary School Journal*, 111 no. 3 (2011): 430–454; Misty Sailors and Nancy L. Shanklin, "Growing Evidence to Support Coaching in Literacy and Mathematics," *The Elementary School Journal* 111, no. 1 (2010): 1–6.
3. Lynsey K. Gibbons and Paul Cobb, "Content-Focused Coaching Practices: Five Key Practices," *Elementary School Journal*, 117, no. 2 (2016): 237–260.
4. Deborah L. Ball and David K. Cohen, "Developing Practice, Developing Practitioners: Toward a Practice-based Theory of Professional Education," in *Teaching as the Learning Profession: Handbook of Policy and Practice*, eds. Linda Darling-Hammond and Gary Sykes (San Francisco: Jossey Bass, 1999), 3-32; Pam Grossman, et al., "Teaching Practice: A Cross-Professional Perspective," *Teachers College Record* 111, no. 9 (2009): 2055-2100; Peter Smagorinsky, Leslie S. Cook and Tara S. Johnson, "The Twisting Path of Concept Development in Learning to Teach," *Teachers College Record* 105, no. 8 (2003): 1399–1436.
5. Matthew A. Kraft, David Blazar and Dylan Hogan, *The Effect of Teacher Coaching on Instruction and Achievement: A Meta-Aanalysis of the Causal Evidence* (Brown University Working Paper, 2016).
6. In Jennifer York-Barr and Karen Duke, "What Do We Know About Teacher Leadership? Findings from Two Decades of Scholarship," *Review of Educational Research* 74, no. 3 (2004): 255–316, York-Barr and Duke argue that teacher leadership has been broadly and inconsistently defined within the literature: It has been variously understood as teachers' participation in: (1) school organization and management duties (i.e., attending site-based management meetings); (2) supporting teacher learning and instructional improvement (i.e., leading teachers' collaborative time, such as "professional learning communities"); and (3) in supporting the "reculturation" of schools, so that initiatives intended to support teacher learning might take root. In this chapter, we are most interested in the second definition of teacher leadership, since, overall, our work seeks to understand how to support instructional improvement at scale.
7. Susan C. Cantrell and Hannah K. Hughes, "Teacher Efficacy and Content Literacy Implementation: An Exploration of the Effects of Extended Professional Development with Coaching," *Journal of Literacy Research* 40 (2008): 95–127; Laura M. Desimone and Katie Pak, "Instructional Coaching as High-Quality Professional Development," *Theory Into Practice*, 56, no. 1 (2017): 3–12; Jim Knight, "Instructional COACHING," *School Administrator* 63, no. 4 (2006): 36–40; Barbara Neufeld and Dana Roper, *Coaching: A Strategy for Developing Instructional Capacity, Promises and Practicalities* (Washington, DC: Aspen Institute Program on Education and Annenberg Institute for School Reform, 2003); Susan B. Neuman and Tanya S. Wright, "Promoting Language and Literacy Development for Early Childhood Educators: A Mixed-Methods Study of Coursework and Coaching," *Elementary School Journal* 111, no. 1 (2010): 63–86; Christine M. Neumerski, "Rethinking Instructional Leadership, a Review: What Do We Know About Principal, Teacher and Coach Instructional Leadership and Where Should We Go From Here?" *Educational Administration Quarterly* 49, no. 2 (2013): 310–347; Sailors and Shanklin, "Growing Evidence"; Beverly Showers and Bruce Joyce, "The Evolution of Peer Coaching," *Educational*

Leadership 53, no. 6 (1996): 12–16; Sharon Walpole et al., "The Relationships Between Coaching and Instruction in the Primary Grades: Evidence from High-Poverty Schools," *Elementary School Journal* 111, no. 1 (2010): 115–140.

8. Chrysan Gallucci et al., "Instructional Coaching: Building Theory about the Role and Organizational Support for Professional Learning," *American Educational Research Journal* 47, no. 4 (2010): 919–963.
9. Judith Warren Little, "The Persistence of Privacy: Autonomy and Initiative in Teachers' Professional Relations," *Teachers College Record* 91, no. 4 (1990): 509–536; Dan C. Lortie, *Schoolteacher* (Chicago: University of Chicago Press, 1975); York-Barr and Duke, "What Do We Know."
10. Theresa Deussen et al., *"Coach" Can Mean Many Things: Five Categories of Literacy Coaches in Reading First* (Issues & Answers Report, REL 2007-No. 005) (Washington, DC: U.S. Department of Education, Institute of Education Sciences, National Center for Education Evaluation and Regional Assistance, Regional Educational Laboratory Northwest, 2007); Susan M. Poglinco et al., *The Heart of the Matter: The Coaching Model in America's Choice Schools* (Philadelphia: Consortium for Policy Research in Education, University of Pennsylvania, 2003).
11. Kraft, Blazer and Hogan, "Effect of Teacher."
12. Paul Cobb and Kara Jackson, "Towards an Empirically Grounded Theory of Action for Improving the Quality of Mathematics Teaching at Scale," *Mathematics Teacher Education and Development* 13, no. 2 (2011): 6–33.
13. York-Barr and Duke, "What Do We Know."
14. Lynsey Gibbons and Paul Cobb, "Focusing on Teacher Learning Opportunities to Identify Potentially Productive Coaching Activities," *Journal of Teacher Education* (2017), doi.10.1177/0022487117702579
15. By *investigating* instruction, we mean taking steps to better understand the rationale for and enactment of particular instructional practices. This may include reading about a practice, observing it (either live or on video), and/or discussing the practice with others. *Enacting* instruction refers to actual attempts to try a practice, initially with other teachers and then with small groups of students and with a whole class. For more information, see Morva McDonald, Elham Kazemi and Sarah S. Kavanagh, "Core Practices and Pedagogies of Teacher Education: A Call for a Common Language and Collective Activity," *Journal of Teacher Education* 64, no. 5 (2013): 378–386.
16. Gibbons and Cobb, "Focusing on Teacher Learning." Gibbons and Cobb's original manuscript did not include the coaching cycle. In this chapter, we are updating those findings based on recent research from Jennifer Lin Russell, et al., "Learning to Leverage Coaching for Instructional Improvement at Scale: Adaptive Integration in the TN Mathematics Coaching Project," *American Educational Research Journal* (under review).
17. Neuman et al., "Promoting Language."
18. Ibid.
19. Gibbons and Cobb, "Focusing on Teacher Learning".
20. Sharon Feiman-Nemser, "Helping Novices Learn to Teach: Lessons From An Exemplary Support Teacher," *Journal of Teacher Education* 52, no. 1 (2001): 17–30, p. 24.
21. Gibbons and Cobb, "Focusing on Teacher Learning."

22. Charlotte Dunlap, "Examining How School Settings Support Teachers' Improvement of Their Classroom Instruction" (PhD diss., Vanderbilt University, 2016).
23. Gibbons and Cobb, "Content-Focused Coaching"; See also Chapter Eleven.
24. Gibbons and Cobb, "Focusing on Teacher Learning."
25. Jean Lave and Etienne Wenger, *Situated Learning: Legitimate Peripheral Participation* (Cambridge, UK: Cambridge University Press, 1991).
26. Gibbons and Cobb, "Focusing on Teacher Learning."
27. Ibid.
28. Laura M. Desimone and Kate Pak, "Instructional Coaching as High Quality Professional Development. Theory into Practice," *Theory into Practice* 56, no. 1 (2017): 3–12; Russell et al., "Learning to Leverage."
29. Grossman et al., "Teaching Practice."
30. This finding about difficulties implementing the full coaching cycle is consistent with other findings in the literature. See Rita M. Bean et al., "Coaches and Coaching in Reading First Schools: A Reality Check," *The Elementary School Journal*, 111 no. 1 (2010): 87–114.
31. Gibbons and Cobb, "Focusing on Teacher Learning".
32. Ibid.
33. Ibid.
34. David K. Cohen, *Teaching and Its Predicaments.* (Cambridge, MA: Harvard University Press, 2011); Charles Munter, "Developing Visions of High-Quality Mathematics Instruction," *Journal for Research in Mathematics Education* 45, no. 5 (2014): 584–635.
35. Hilda Borko, Karen Koellner and Jennifer Jacobs, "Meeting the Challenges of Scale: The Importance of Preparing Professional Development Leaders," *Teachers College Record*, March 04, 2011. Retrieved from http://www.tcrecord.org ID Number: 16358.
36. Cohen, *Teaching and Its Predicaments*; Ron Ritchhart, Mark Church and Karin Morrison, *Making Thinking Visible: How to Promote Engagement, Understanding, and Independence for All Learners* (San Francisco: Jossey-Bass, 2011).
37. Gibbons and Cobb, "Focusing on Teacher Learning"; Ilana S. Horn, "Fast Kids, Slow Kids, Lazy Kids: Framing the Mismatch Problem in Mathematics Teachers' Conversations," *Journal of the Learning Sciences* 16, no. 1 (2007): 37–79.
38. Gibbons and Cobb, "Focusing on Teacher Learning"; Meilan Zhang et al., "Understanding Affordances and Challenges of Three Types of Video for Teacher Professional Development," *Teaching and Teacher Education* 27, no. 2 (2011): 454–462.
39. Gibbons and Cobb, "Focusing on Teacher Learning"; Elizabeth A. van Es and Miriam G. Sherin, "The Influence of Video Clubs on Teachers' Thinking and Practice," *Journal of Mathematics Teacher Education,* 13 (2010): 155–176.
40. Anne K. Morris and James Hiebert, "Creating Shared Instructional Products: An Alternative Approach to Improving Teaching," *Educational Researcher* 40, no. 1 (2011): 5–14.
41. Lynn C. Hart, Alice S. Alston and Aki Murata, *Lesson Study Research and Practice in Mathematics Education* (New York: Springer, 2011); Catherine Lewis, Rebecca Perry and Aki Murata, "How Should Research Contribute to Instructional Improvement? The Case of Lesson Study," *Educational Researcher* 35, no. 3 (2006): 3–14.
42. Lynsey K. Gibbons, "Examining Mathematics Coaching Practices that Help Develop School-wide Professional Learning," in *Elementary Mathematics Specialists,*

eds. Maggie B. McGatha and Nicole R. Rigelman (Charlotte, NC: Association of Mathematics Teacher Educators, 2017), 167–182; Elham Kazemi et al. "Math Labs: Designing High Quality School-Embedded Math Professional Learning" (paper presented at the annual conference of the *Association of Mathematics Teachers Educators*, Irvine, CA., February 6–8, 2014).

43. Gibbons and Cobb, "Focusing on Teacher Learning."
44. Gina Biancarosa, Anthony S. Bryk and Emily R. Dexter, "Assessing the Value-Added Effects of Literacy Collaborative Professional Development on Student Learning," *The Elementary School Journal* 111, no. 1 (2010): 7–34; Gallucci et al., "Instructional Coaching"; Neufeld et al., *Coaching*.
45. Cynthia Coburn and Jennifer L. Russell, "District Policy and Teachers' Social Networks," *Educational Evaluation and Policy Analysis* 30, no. 3 (2008): 203–235; Rebekah Elliott et al., "Conceptualizing the Work of Leading Mathematical Tasks in Professional Development," *Journal of Teacher Education* 60 (2009): 364–379; Milbrey W. McLaughlin and Joan E. Talbert, *Building School-Based Teacher Learning Communities* (New York: Teachers College Press, 2006); Neumerski, "Rethinking"; William R. Penuel et al., "Analyzing Teachers' Professional Interactions in a School as Social Capital: A Social Network Approach," *Teachers College Record* 111, no. 1 (2009): 124–163; Poglinco et al., *The Heart*.
46. Min Sun et al., "Exploring Colleagues' Professional Influence on Mathematics Teachers' Learning," *Teachers College Record,* 116 (2016): 1–30.
47. Penuel et al., "Analyzing Teachers' Professional"; McLaughlin and Talbert, *Building*; Neumerski, "Rethinking"; Elliott, "Conceptualizing"; Coburn and Russell, "District Policy"; Poglinco et al., *The Heart.*
48. Dunlap, "Examining."
49. York-Barr and Duke, "What Do We Know."
50. Ibid; Britnie D. Kane and Brooks Rosenquist, "Understanding District- and School-Level Policies and Expectations Around Instructional Coaching: How They Influence Coaches' Time Engaged in Activities Likely to Support Teachers' Instructional Improvement," *American Educational Research Journal* (under review).
51. Horn et al., "Opportunities for Professional Learning"; Karen S. Louis, Helen M. Marks and Sharon Kruse, "Teachers' Professional Community in Restructuring Schools," *American Educational Research Journal* 33 (1996): 757–798; Pam Grossman, Sam Wineburg and Stephen Woolworth, "Toward a Theory of Teacher Community," *Teachers College Record,* 103, no. 6 (1999): 942–1012; McLaughlin and Talbert, *Building.*
52. Dan C. Lortie, *Schoolteacher.* (Chicago: University of Chicago Press, 1975); Little, "The Persistence."
53. Kane and Rosenquist, "Understanding District- and School-Level Policies."
54. Megan Webster, "A Decomposition of the Practices of High Quality Professional Development Facilitation for Teachers" (PhD diss., McGill University, 2015).
55. Gibbons and Cobb, "Content-Focused Coaching."
56. Cathy A. Toll, *The Literacy Coach's Survival Guide: Essential Questions and Practical Answers.* 2nd edition. (Newark, DE: International Reading Association, 2014); Sharon Walpole and Michael C. McKenna, *The Literacy Coach's Handbook: A Guide to Research-Based Practice.* 2nd edition. (New York: The Guilford Press, 2013).

57. Although Gibbons and Cobb, in "Focusing on Teacher Learning," did not find enough empirical evidence for observing and debriefing classrooms in order to identify this coaching activity as potentially productive, it was included in this analysis because, as we discussed, this coaching activity may be a valuable support for teachers' learning if complemented by other potentially productive coaching activities.
58. Gibbons et al., "Content-Focused Coaching."
59. Gibbons and Cobb, "Focusing on Teacher Learning"; McLaughlin and Talbert, *Building*; Etienne Wenger, *Communities of Practice: Learning, Meaning, and Identity*. (Cambridge: Cambridge University Press, 1998).
60. Ilana S. Horn et al., "A Taxonomy of Instructional Learning Opportunities in Teachers' Workgroup Conversations," *Journal of Teacher Education* 68, no. 1 (2017): 41-54.
61. Webster, "A Decomposition of the Practices"; Ilana S. Horn and Britnie D. Kane, "Opportunities for Professional Learning in Mathematics Teacher Workgroup Conversations: Relationships to Instructional Expertise," *Journal of the Learning Sciences* 24, no. 3 (2015): 373-418; See also Chapter Five.
62. Maryl Gearhart, et al., "Developing Expertise with Classroom Assessment in K-12 Science: Learning to Interpret Student Work. Interim Findings from a 2-year Study," *Educational Assessment* 11 (2006): 237-263; Gibbons and Cobb, "Focusing on Teacher Learning"; Pam Grossman et al., "Toward a Theory of Teacher Community"; Ilana S. Horn, "Learning on the Job: A Situated Account of Teacher Learning in High School Mathematics Departments," *Cognition and Instruction* 23, no. 2 (2005): 207-236; Horn, et al., "Taxonomy of Instructional Learning"; Britnie D. Kane, "Facilitating Sensemaking in Teacher Workgroups: Concept Development Through Representations of Practice," *Journal of the Learning Sciences* (under review); Elham Kazemi, and Megan L. Franke, "Teacher Learning in Mathematics: Using Student Work to Promote Collective Inquiry," *Journal of Mathematics Teacher Education* 7 (2004): 203-235; Judith Warren Little et al., "Looking at Student Work for Teacher Learning, Teacher Community, and School Reform," *Phi Delta Kappan* 85 (2003): 184-192; Miriam G. Sherin, "New Perspectives on the Role of Video in Teacher Education," In *Advances in Research on Teaching: Using Video in Teacher Education*, ed. Jere Brophy (Oxford, UK: Elsevier, 2004), 1-27; Miriam G. Sherin and Sandra Y. Han, "Teacher Learning in the Context of Video Club," *Teaching and Teacher Education* 20 (2004): 163-183; Jessica Thompson, et al., "Examining Student Work: Evidence-based Learning for Students and Teachers," *The Science Teacher* 76, no. 8 (2009): 48-52.
63. Elliott et al., "Conceptualizing the Work"; Ann C. Howe and Harriett S. Stubbs, "From Science Teacher to Teacher Leader: Leadership Development as Meaning Making in Community of Practice," *Science Teacher Education* 87 (2003): 281-297; Ann Lieberman and Maureen Grolnick, "Networks and Reform in American Education," *Teachers College Record* 98, no. 1 (1996): 7-45.
64. Gibbons and Cobb, "Focusing on Teacher Learning".
65. Deussen et al., *"Coach" Can Mean*; Gallucci et al., "Instructional Coaching"; Neumerski, "Rethinking"; Poglinco et al., *The Heart*; York-Barr and Duke, "What Do We Know."
66. Rita M. Bean, et al., "Coaches and Coaching in Reading First Schools: A Reality Check," *The Elementary School Journal* 111, no. 1 (2010): 87-114; Deussen et al., *Coach Can Mean.*

67. Melinda M. Mangin and KariLonnie Dunsmore, "How the Framing of Instructional Coaching as a Lever for System or Individual Reform Influences the Enactment of Coaching," *Education Administration Quarterly* 51, no. 2 (2015): 179–213.
68. Poglinco et al., *The Heart.*
69. Bean et al., "Coaches and Coaching"; Deussen et al., *"Coach" Can Mean;"* Neumerski, "Rethinking."
70. Kane and Rosenquist, "Understanding District- and School-Level Policies."
71. Lynsey K. Gibbons, Anne Garrison Wilhelm and Paul Cobb, "Coordinating Leadership Supports for Teachers' Instructional Improvement" (under review), *Journal of Research on Organization in Education*; Lindsay C. Matsumura et al., "Leadership for Literacy Coaching: The Principal's Role in Launching a New Coaching Program," *Educational Administration Quarterly* 45 no. 5 (2009): 655–693; Neumerski, "Rethinking."
72. Gibbons et al., "Coordinating Leadership."
73. Kane and Rosenquist, "Understanding District- and School-Level Policies."
74. Ibid.
75. Here, we report on our finding that, in one of our partner districts during years three and four of our study, coaches hired by principals to work in a single school spent 42% of their time engaging teachers in activities related to instructional improvement. Careful readers will note that the same percentage was reported earlier, as the average amount of time that all coaches across all participating districts in our six years of survey data collection claimed to spend engaging teachers in work related to instruction. It is a coincidence that both of these averages are 42%.
76. For examples of research that suggests principals act as a barrier to coaching and teacher leadership, see Neumerski, "Rethinking" and York-Barr and Duke, "What Do We Know." Poglinco et al., in *The Heart*, noted that the coaches in their studies did not believe that principals had a clear idea of how they might work with them productively.
77. Kane and Rosenquist, "Understanding District- and School-Level Policies."
78. Chrysan Gallucci et al., "Instructional Coaching"; Poglinco et al., *The Heart.*
79. Kara Jackson, et al., "Investigating the Development of Mathematics Leaders' Capacity to Support Teachers' Learning on a Large Scale," *ZDM Mathematics Education* 47 (2015): 93–104; See also Chapter Four.
80. Horn, "Fast Kids, Slow Kids"; see also chapter 5 of this volume. For an exception, see Russell et al., "Learning to Leverage."

CHAPTER 8

1. We would like to thank Min Sun, Jessica Rigby, Tom Smith, Dan Berebitsky, Christy Andrews-Larson, and Adrian Larbi-Cherif for their substantial contributions to this line of inquiry focused on teachers' advice networks in the MIST project.
2. Alan J. Daly et al., "Relationships in Reform: The Role of Teachers' Social Networks," *Journal of Educational Administration* 48, no. 3 (2010): 359–91.
3. Charles E. Bidwell and Jeffrey Y. Yasumoto, "The Collegial Focus: Teaching Fields, Collegial Relationships, and Instructional Practice in American High Schools," *Sociology of Education* 72, no. 4 (1999): 234-56; Kenneth A. Frank, Yong Zhao and Kathryn Borman, "Social Capital and the Diffusion of Innovations within

Organizations: The Case of Computer Technology in Schools," *Sociology of Education* 77 (2004): 148–71, doi:10.1177/003804070407700203; Leigh Mesler Parise and James P. Spillane, "Teacher Learning and Instructional Change: How Formal and on-the-Job Learning Opportunities Predict Change in Elementary School Teachers' Practice," *The Elementary School Journal* 110, no. 3 (2010): 323–46; William R. Penuel et al., "Analyzing Teachers' Professional Interactions in a School as Social Capital: A Social Network Approach," *Teachers College Record* 111, no. 1 (2009): 124–63.

4. Frank et al. "Social capital"; Kenneth A. Frank et al., "Focus, Fiddle, and Friends: Knowledge to Perform Complex Tasks," *Sociology of Education* 84, no. 2 (2011): 137–156, doi:10.1177/0038040711401812
5. Mark A. Smylie and Andrea E Evans, "Social Capital and the Problem of Implementation," in *New Directions in Education Policy: Confronting Complexity*, ed. Meredith I. Honig (New York: SUNY Press, 2006),187–208; James P. Spillane, Brian J. Reiser and Louis M. Gomez, "Policy Implementation and Cognition," in *New Directions in Educational Policy Implementation*, ed. Meredith I. Honig (New York: SUNY Press, 2006), 47–64.
6. Cynthia E. Coburn, Linda Choi and Willow Mata, "I Would Go to Her Because Her Mind Is Math": Network Formation in the Context of a District-Based Mathematics Reform," in *Social Network Theory and Educational Change*, ed. Alan J. Daly (Cambridge, Massachusetts: Harvard Education Press, 2010), 33–50; Anne Garrison Wilhelm et al., "Selecting Expertise in Context: Middle School Mathematics Teachers' Selection of New Sources of Instructional Advice," *American Educational Research Journal* 53, no. 3 (2016): 456–91.
7. See for example Bidwell and Yasumoto, "The Collegial Focus"; Penuel et al., "Analyzing Teachers' Professional Interactions."
8. Min Sun et al., "Exploring Colleagues' Professional Influence on Mathematics Teachers' Learning," *Teachers College Record* 116, no. 6 (2014): 1–30.
9. Cynthia E. Coburn and Jennifer Lin Russell, "District Policy and Teacher's Social Networks," *Educational Evaluation and Policy Analysis* 30, no. 3 (2008): 203–35.
10. Daniel Berebitsky and Christine Andrews-Larson, "Teacher Advice Seeking: Relating Centrality and Expertise in Middle School Mathematics Social Networks," *Teachers College Record* 119, no. 10 (2017); Anne Garrsion Wilhelm et al., "Selecting Expertise in Context."
11. Lynsey K. Gibbons, Anne Garrison Wilhelm and Paul Cobb, "Coordinating Leadership Supports for Teachers' Instructional Improvement" (under review), *Journal of Research on Organization in Education.*
12. Wilhelm et al., "Selecting Expertise in Context."
13. I-Chien Chen et al., "From Conversation to Collaboration: How the Quality of Teacher Workgroup Meetings Influences Social Networks," (paper presented at the annual meeting of the *American Education Research Association*, San Antonio, TX, April 27-May 1, 2017.)
14. Ilana Seidel Horn, "Teaching Replays, Rehearsals, and Re-Visions." *Teachers College Record* 112, no. 1 (2010): 225–59; Ilana Seidel Horn et al, "A Taxonomy of Instructional Learning Opportunities in Teachers' Workgroup Conversations," *Journal of Teacher Education* 68, no. 1 (2016): 41–54.
15. Ibid.

16. Chen et al., "From Conversation to Collaboration."
17. Berebitsky and Andrews-Larson, "Teacher Advice Seeking."
18. Gibbons, Wilhelm and Cobb, "Coordinating Leadership Supports".
19. This was shown to be the case with MIST data in Berebitsky and Andrews-Larson, "Teacher Advice Seeking." Other studies that show that when people are assigned to formal roles (e.g., instructional coach, department chair), the chances that they will be sought out for instructional advice increases include Nienke M. Moolenaar et al., "Social Forces in School Teams: How Demographic Composition Affects Social Relationships," in *Interpersonal Relationships in Education: From Theory to Practice*, eds. David Zandvliet et al. (Rotterdam, The Netherlands: Sense Publishers, 2014), 159–181; James P. Spillane and Chong Min Kim, "An Exploratory Analysis of Formal School Leaders' Positioning in Instructional Advice and Information Networks in Elementary Schools," *American Journal of Education* 119, no. 1 (2012): 73–102; James P. Spillane, Chong Min Kim, and Kenneth A. Frank, "Instructional Advice and Information Providing and Receiving Behavior in Elementary Schools: Exploring Tie Formation as a Building Block in Social Capital Development," *American Educational Research Journal* 49, no. 6 (2012): 1112–45; James P. Spillane, Megan Hopkins and Tracy Sweet, "Intra- and Interschool Interactions About Instruction: Exploring the Conditions for Social Capital Development," *American Journal of Education* 122, no. 1 (2015): 71–110.
20. Wilhelm et al., "Selecting Expertise In Context"; Kenneth A. Frank and Yong Zhao, "Subgroups as Meso-Level Entities in the Social Organization of Schools," in *The Social Organization of Schooling*, eds. Larry V. Hedges and Barabara L. Schneider (New York: Russell Sage Foundation, 2005), 200–224; Moolenaar et al., "Social Forces in School Teams"; William R. Penuel et al., "The Alignment of the Informal and Formal Organizational Supports for Reform: Implications for Improving Teaching in Schools," *Educational Administration Quarterly* 46, no. 1 (2010): 57–95; Spillane et al., "Intra- and Interschool Interaction"; Spillane et al., "Instructional Advice and Information Providing."
21. Jessica G. Rigby, Christine Andrews-Larson and I-Chien Chen, "Teachers' Learning Opportunities about Teaching Mathematics: A Longitudinal Case Study of Administrators' Influence," *Journal of Educational Change* (under review).
22. Wilhelm et al., "Selecting Expertise in Context."
23. Spillane et al., "Instructional Advice and Information Providing"; Sun et al., "Exploring Colleagues' Professional Influence."
24. S. P. Borgatti, *NetDraw Software for Network Visualization*. (Lexington, KY: Analytic Technologies, 2012).
25. Coburn and Russell, "District Policy and Teacher's Social Networks"; Cynthia E. Coburn, Willow S. Mata and Linda Choi, "The Embeddedness of Teachers' Social Networks: Evidence from a Study of Mathematics Reform," *Sociology of Education* 86, no. 4 (2013): 311–342.
26. For an example of normative influence of teacher networks see Kenneth A. Frank, Chong M. Kim and Dale Belman, "Utility Theory, Social Networks, and Teacher Decision Making," in *Social Network Theory and Educational Change*, ed. Alan J. Daly (Cambridge: Harvard University Press, 2010), 223–242.

CHAPTER 9

1. Mary Kay Stein and Kim Gooyeon, "The Role of Mathematics Curriculum Materials in Large-Scale Urban Reform," in *Mathematics Teachers at Work: Connecting Curriculum Materials and Classroom Instruction,* eds. Janine T. Remillard, Beth A. Herbel-Eisenmann and Gwendolyn M. Lloyd (Abingdon, UK: Routledge, 2009), 37-55; Mary Kay Stein, Janine Remillard, and Margaret S. Smith, "How Curriculum Influences Student Learning," *Second Handbook of Research on Mathematics Teaching and Learning,* ed. Frank K. Lester, Jr. (Charlotte: Information Age Publishing, 2007), 319–370.
2. Rachel Collopy, "Curriculum Materials as a Professional Development Tool: How a Mathematics Textbook Affected Two Teachers' Learning," *The Elementary School Journal* 103, no. 3 (2003): 287–311; Lynn T. Goldsmith, Helen M. Doerr and Catherine C. Lewis, "Mathematics Teachers' Learning: A Conceptual Framework and Synthesis of Research," *Journal of Mathematics Teacher Education* 17, no. 1 (2014): 5–36; Janine Remillard and Martha B. Bryans, "Teachers' Orientations toward Mathematics Curriculum Materials: Implications for Teacher Learning," *Journal for Research in Mathematics Education* 35, no. 5 (2004): 352–388.
3. Glenda Lappan et al., *Connected Mathematics Project 2* (Boston: Pearson, 2009).
4. Amy Crosson et al., "The Instructional Quality Assessment as a Professional Development Tool" (paper presented at the annual meeting of the *American Educational Research Association,* San Diego, CA, April 12–16, 2004).
5. Mary Kay Stein, Barbara W. Grover and Marjorie Henningsen, "Building Student Capacity for Mathematical Thinking and Reasoning: An Analysis of Mathematical Tasks used in Reform Classrooms," *American Educational Research Journal* 33, no. 2 (1996): 455–488; James E. Tarr, et al., "From the Written to the Enacted Curricula: The Intermediary Role of Middle School Mathematics Teachers in Shaping Students' Opportunity to Learn," *School Science and Mathematics* 106, no. 4 (2006): 191–201; James E. Tarr et al., "The Impact of Middle-grades Mathematics Curricula and the Classroom Learning Environment on Student Achievement," *Journal for Research in Mathematics Education* 39, no. 3 (2008): 247–280.
6. Brooks A. Rosenquist, Anne G. Wilhelm and Thomas M. Smith, "The IQA Observational Rubric and Teacher Value-added: Looking for Nonlinearities to Inform Practice and Policy" (paper presented at the annual meeting of the *American Educational Research Association*, Philadelphia, Pennsylvania, April 3–7, 2014).
7. Anne G. Wilhelm. "Mathematics Teachers' Enactment of Cognitively Demanding Tasks: Investigating Links to Teachers' Knowledge and Conceptions," *Journal for Research in Mathematics Education*, 45 no. 5 (2014): 636–674.
8. Fred M. Newmann et al., "Instructional Program Coherence: What It Is and Why It Should Guide School Improvement Policy," *Educational Evaluation and Policy Analysis* 23 (2001): 297-321.
9. Mollie Appelgate and Brooks A. Rosenquist, "As *Common Core* Takes Hold: Changes in Teachers' Mathematics Curriculum Use" (paper presented at the *National Council of Teachers of Mathematics Research Conference,* San Francisco, California, April 11–13, 2016).
10. Stein et al., "How Curriculum Influences."
11. Tarr et al., "From the Written"; Tarr et al., "The Impact of Middle-grades."

CHAPTER 10

1. Paul Black and Dylan Wiliam, "Inside the Black Box: Raising Standards Through Classroom Assessment," *Phi Delta Kappan* 91, no. 1 (1998): 81–90; Carol Anne Dwyer, "Assessment and Classroom Learning: Theory and Practice," *Assessment in Education* 5, no. 1 (1998): 131–137; James W. Pellegrino, Naomi Chudowsky and Robert Glaser, *Knowing What Students Know: The Science and Design of Education Assessment* (Washington DC: National Academy Press, 2001); W. James Popham, *Test Better, Teacher Better: The Instruction Role of Assessment* (Alexandria, VA: Association for Supervision Curriculum Development, 2003).
2. Brette Garner and Ilana S. Horn, "Using Standardized Test Data as a Starting Point for Inquiry: A Case for Thoughtful Compliance," (in press) in *Teachers' Data Use: Cases of Promising Practice*, eds. Nicole Barnes and Helenrose Fives (New York City: Routledge).
3. Jennifer Booher-Jennings, "Below the Bubble: 'Educational Triage' and the Texas Accountability System," *American Educational Research Journal* 42, no. 2 (2005): 231–268; John B. Diamond and Kristy Cooper, "The Uses of Testing Data in Urban Elementary Schools: Some Lessons from Chicago," *Evidence and Decision Making* 106, no. 1 (2007): 241–263; Ilana S. Horn, "Accountability as a Design for Teacher Learning: Sensemaking About Mathematics and Equity in the NCLB Era," *Urban Education* (2016) DOI:10.1177/0042085916646625; Jennifer L. Jennings and Jonathan Marc Bearak, "'Teaching to the Test' in the NCLB Era: How Test Predictability Affects Our Understanding of Student Performance," *Educational Researcher* 43, no. 8 (2014): 381–389.
4. Kun Yuan and Vi-Nhuan Le, *Estimating the Percentage of Students who were Tested on Cognitively Demanding Items Through the State Achievement Tests* (Santa Monica, CA: RAND Corporation, 2012), http://www.rand.org/content/dam/rand/pubs/working_papers/2012/RAND_WR967.pdf
5. Jennings and Bearak, "Teaching to the Test."
6. Brooks A. Rosenquist, Charlotte Sharpe and Emily Kern, "Marshaling Evidence for Increased Test Rigor for Career and College Ready Standards-aligned State Assessments" (working paper, MIST project, Vanderbilt University, 2016).
7. Black and Wiliam, "Inside the Black Box."
8. The FALs were implemented as part of the Math Design Collaborative, http://map.mathshell.org/lessons.php
9. Ilana S. Horn, Britnie D. Kane and Jonee Wilson, "Making Sense of Student Performance Data," *American Educational Research Journal* 52, no. 2 (2015): 208–242.
10. Jason Brasel et al., "Getting to the Why and How," *Educational Leadership* 73, no. 3 (2015).
11. Garner and Horn, "Using Standardized Test Data."
12. Elham Kazemi et al., "Doing Mathematics in Professional Development: Working with Leaders to Cultivate Mathematically Rich Teacher Learning Environments," in *Association of Mathematics Teacher Educators Monograph VI: Scholarly Practices and Inquiry in the Preparation of Mathematics Teachers* eds. Konrad Krainer and Terry Woods (San Diego, CA: Association of Mathematics Teacher Educators, 2009); Magdalene Lampert et al., "Using Designed Instructional Activities to Enable Novices to Manage Ambitious Mathematics Teaching," *Instructional Explanations in the Disciplines*, eds. Mary Kay Stein and Linda Kucan (New York: Springer, 2010), 129–141.
13. Alice Huguet, Julie A. Marsh and Caitlin Farrell, "Building Teachers' Data-use

Capacity: Insights from Strong and Developing Coaches," *Education Policy Analysis Archives* 22, no. 52 (2014): 1–31.

14. Ilana S. Horn, Brette Garner, Britnie D. Kane Jason Brasel, "A Taxonomy of Instructional Learning Opportunities in Teachers' Workgroup Conversations," *Journal of Teacher Education* 68, no. 1 (2017): 41–54.
15. Christopher L. Miller, "Accountability Policy Implementation and the Case of Smaller School District Capacity: Three Contrasting Cases that Examine the Flow and Use of NCLB Accountability Data," *Leadership and Policy in Schools* 9, no. 4 (2010): 384–420.
16. Vicki Park, Alan J. Daly and Alison W. Guerra, "Strategic Framing: How Leaders Craft the Meaning of Data Use for Equity and Learning," *Educational Policy* 27, no. 4 (2012): 645–675.

CHAPTER 11

1. Acknowledgments: We would like to thank Kara Jackson, Paul Cobb, and Rebecca Schmidt, as well as our teacher participants and district leader partners, for their substantial contributions to the line of inquiry focused on supports for struggling students in the MIST project.
2. By "currently struggling students," we are referring to students who were the target of additional instruction. Students were typically identified as low performing on the basis of state assessments. They generally were not the same population of students as those identified as eligible to receive special education services. In our use of the term, we are not suggesting that *struggling* is a static trait of the students. Rather we take the perspective that their struggle is likely a product of having not been supported to participate substantially in rigorous mathematical activity over the course of many years.
3. June Mark, Josephine Louie, and Mary Fries, "Supporting Students to Succeed in Algebra: Strategies and Resources" (paper presented at the annual meeting of the National Council of Supervisors of Mathematics, Philadelphia, Pennsylvania, April 23–25, 2012); Claire Durwood, Emily Krone and Christopher Mazzeo, *Are Two Algebra Classes Better than One? The Effects of Double-dose Instruction in Chicago* (Chicago: Consortium on Chicago School Research, 2010).
4. Mark et al., "Supporting Students."
5. Linda Darling-Hammond, *The Flat Earth and Education: How America's Commitment to Equity will Determine our Future* (New York: Teachers College Press, 2010).
6. Mark et al., "Supporting Students."
7. Attending to the quality of supplemental instruction was not initially in our scope of work, and therefore, we did not budget for resources to collect video recordings of supplemental instruction. However, we quickly recognized the prominence of supplemental classes, and asked about the goals of supplemental instruction, how supplemental classes were organized, the curriculum that was used, and the challenges teachers encountered in our interviews with central office leaders, school leaders, and teachers.
8. Rebecca A. Schmidt, "Unpacking Tracking: The Role of Instruction, Teacher Beliefs and Supplemental Courses in the Relationship between Tracking and Student Achievement" (PhD diss., Vanderbilt University, 2013).
9. Ibid.

10. Schmidt hypothesized that this could be due to the re-distribution of existing resources -without having been provided additional resources. For example, as we describe later in the chapter, for the most part, teachers were asked to create and teach a new course, without curricular resources or professional development. This could have led them to use their planning time to pull together resources for the supplemental class, as opposed to focusing on planning for their other mainstream classes.
11. Durwood, et al., *Are Two Algebra Classes*; Takako Nomi and Elaine Allensworth, "'Double-dose Algebra as an Alternative Strategy to Remediation: Effects on Students' Academic Outcomes," *Journal of Research on Educational Effectiveness* 2, no. 2 (2009): 111–148.
12. Schmidt, "Unpacking Tracking."
13. Ibid.
14. Ibid.
15. Ibid.
16. Kris D. Gutiérrez, P. Zitlali Morales and Danny C. Martinez, "Re-mediating Literacy: Culture, Difference, and Learning for Students from Nondominant Communities," *Review of Research in Education* 33, no. 1 (2009): 212–245, doi:10.3102/0091732X08328267.
17. Heather C. Hill, Deborah L. Ball and Stephen G. Schilling, "Unpacking Pedagogical Content Knowledge: Conceptualizing and Measuring Teachers' Topic-specific Knowledge of Students," *Journal for Research in Mathematics Education,* 39, no. 4 (2008): 372–400.
18. Elisabeth G. Cohen and Rachel A. Lotan, *Designing Groupwork: Strategies for the Heterogeneous Classroom. 3rd Edition*. (New York and London: Columbia University, Teachers College Press, 2014), 70.
19. *Principles to Actions: Ensuring Mathematical Success for All* (Reston, VA: National Council of Teachers of Mathematics, 2014).

CHAPTER 12

1. We would like to thank Jessica G. Rigby, Karin Katterfeld, and Brooks Rosenquist for their substantial contributions to this line of inquiry.
2. Viviane M. J. Robinson, Claire A. Lloyd and Kenneth J. Rowe, "The Impact of Leadership on Student Outcomes: An Analysis of the Differential Effects of Leadership Types," *Educational Administration Quarterly* 44, no. 5 (2008): 635–674; Cynthia E. Coburn, "Shaping Teacher Sensemaking: School Leaders and the Enactment of Reading Policy," *Educational Policy* 19, no. 3 (2005): 476–509.
3. Peter Youngs and M. Bruce King, "Principal Leadership for Professional Development to Build School Capacity," *Educational Administration Quarterly* 38, no. 5 (2002): 643–670.
4. Reform Support Network, "Using Observations to Improve Teacher Practice How States Can Build Meaningful Observation Systems," (2015) https://www2.ed.gov/about/inits/ed/implementation-support-unit/tech-assist/usingobservationsto Improveteacherpractice.pdf; Bruce S. Cooper, Patricia A. L. Ehrensal, Matthew Bromme, "School-level Politics and Professional Development: Traps in Evaluating the Quality of Practicing Teachers," *Educational Policy* 19, no. 1 (2005); 112–125;

Carolyn J. Downey Betty E. Steffy, Fenwick W. English, Larry E. Frase and William K. Poston, *The Three-minute Classroom Walkthrough,* (Thousand Oaks, CA: Corwin, 2004); Elaine Fink and Lauren B. Resnick, "Developing Principals as Instructional Leaders," *Phi Delta Kappan* 82, no. 8 (2001): 598–606.

5. Melissa D. Boston et al., "Investigating How to Support Principals as Instructional Leaders in Mathematics," *Journal of Research on Leadership Education* (2016), doi: 10.1177/1942775116640254.
6. Jessica G. Rigby et al., "Administrator Observation and Feedback: Does It Lead Toward Improvement in Inquiry-Oriented Math Instruction?" *Educational Administration Quarterly* (2017): 1–42, doi: 10.1177/0013161X16687006.
7. Charlotte J. Dunlap, "Principals of Productive Teacher Collaboration: An Integrative Conceptual Analysis of the Teacher Learning and Instructional Leadership Literatures" (Major Area Paper, Vanderbilt University, 2015).
8. Britnie D. Kane and Brooks Rosenquist, "Understanding District- and School-level Policies and Expectations around Instructional Coaching: How They Influence Coaches' Time Engaged in Activities Likely to Support Teachers' Instructional Improvement," (under review) *American Educational Research Journal.*
9. Melinda M. Mangin, "Literacy Coach Role Implementation: How District Context Influences Reform Efforts," *Educational Administration Quarterly* 45, no. 5 (2009): 759–792; Lindsay Clare Matsumura, et al., "Leadership for Literacy Coaching: The Principal's Role in Launching a New Coaching Program," *Educational Administration Quarterly* 45, no. 5 (2009): 655–693.
10. Mangin, "Literacy Coach Role Implementation"; Sebastian Wren and Diana Vallejo. "Effective Collaboration between Instructional Coaches and Principals," (2009). http://www.balancedreading.com/Wren_&_Vallejo_Coach_Principal_Relatinships.pdf.
11. Lynsey K. Gibbons, Anne Garrison Wilhelm and Paul Cobb, "Coordinating Leadership Supports for Teachers' Instructional Improvement" (under review), *Journal of Research on Organization Education.*
12. Jessica G. Rigby, Christine Andrews-Larson and I-Chien Chen, "Teachers' Learning Opportunities about Teaching Mathematics: A Longitudinal Case Study of Administrators' Influence," (under review) *Journal of Educational Change.*
13. Rigby et al., "Teachers' Learning Opportunities."
14. Adrian Larbi-Cherif, "Investigating Relationships between Understanding of Inquiry Mathematics, District Context, and School Context on Principal Instructional Leadership Aimed at Ambitious Instruction" (PhD diss., Vanderbilt University, 2017).
15. Ibid.
16. Gibbons et al., "Coordinating Leadership"; Larbi-Cherif, "Investigating Relationships."
17. Larbi-Cherif, "Investigating Relationships."
18. Ibid.
19. Ibid.
20. Boston et al., "Investigating How."
21. Ibid.
22. Larbi-Cherif, "Investigating Relationships."
23. Ibid.

24. Charlotte J. Dunlap, Megan Webster, Kara Jackson, and Paul Cobb, "Schooling Leaders on the Common Core." *Phi Delta Kappan Common Core Writing Project.* 2014. Accessed August 30, 2017.http://www.kappancommoncore.org/schooling-leaders-on-the-common-core/.
25. Richard F. Elmore, "Leadership as the Practice of Improvement" in *Improving School Leadership, Vol. 2: Case Studies on System Leadership*, eds. David Hopkins, Deborah Nusche and Beatriz Pont (Paris: OECD Publishing, 2008).

CHAPTER 13

1. We would like to thank Erin Henrick, Charlotte Sharpe, and Megan Webster for their substantial contributions to this line of inquiry.
2. John E. Chubb and Terry M. Moe, *Politics, Markets, and America's Schools* (Washington, DC: The Brookings Institution, 1990); Denis P. Doyle and Chester E. Finn, Jr., "American Schools and the Future of Local Control," *Public Interest* 77 (1984): 77–95; Tina Trujillo, "The Reincarnation of the Effective Schools Research: Rethinking the Literature on District Effectiveness," *Journal of Educational Administration* 51, no. 4 (2013): 426–452.
3. Meredith I. Honig, "District Central Offices as Learning Organizations: How Sociocultural and Organizational Learning Theories Elaborate District Central Office Administrators' Participation in Teaching and Learning Improvement Efforts," *American Journal of Education* 114, no. 4 (2008): 627–664; Meredith I. Honig and Cynthia E. Coburn, "Evidence-Based Decision Making in School District Central Offices: Toward a Policy and Research Agenda," *Educational Policy* 22, no. 4 (2008): 578–608. Mary Kay Stein and Cynthia E. Coburn, "Architectures for Learning: A Comparative Analysis of Two Urban School Districts," *American Journal of Education* 114, no. 4 (2008): 583–626; Charles L. Thompson, Gary Sykes and Linda Skrla, *Coherent, Instructionally-Focused District Leadership: Toward a Theoretical Account* (Lansing, MI: The Education Policy Center at Michigan State University, 2008).
4. For exceptions, see, for example, Patricia Burch and James P. Spillane, "How Subjects Matter in District Office Practice: Instructionally Relevant Policy in Urban School District Redesign," *Journal of Educational Change* 6, no. 1 (2005): 51–76; Meredith I. Honig, "District Central Office Leadership as Teaching: How Central Office Administrators Support Principals' Development as Instructional Leaders," *Educational Administration Quarterly* 48, no. 4 (2012): 733–774.
5. Paul Cobb and Kara Jackson, "Analyzing Educational Policies: A Learning Design Perspective," *The Journal of the Learning Sciences* 21, no. 4 (2012): 487–521.
6. Ibid.
7. James E. Tarr et al., "The Impact of Middle-Grades Mathematics Curricula and the Classroom Learning Environment on Student Achievement, *Journal for Research in Mathematics Education* 39, no. 3 (2008): 247–280. See also Julie E. Riordan and Pendred E. Noyce, "The Impact of Two Standards-Based Mathematics Curricula on Student Achievement in Massachusetts," *Journal for Research in Mathematics Education* 32, no. 4 (2001): 368–398; and Alan Schoenfeld, "Making Mathematics Work for All Children: Issues of Standards, Testing, and Equity," *Educational Researcher* 31, no. 1 (2002): 13–25.
8. Cobb and Jackson, "Analyzing Educational Policies." See also chapter 10.

9. Britnie D. Kane and Brooks Rosenquist, "Understanding District- and School-Level Policies and Expectations Around Instructional Coaching: How They Influence Coaches' Time Engaged in Activities Likely to Support Teachers' Instructional Improvement," *American Educational Research Journal* (under review).
10. Cobb and Jackson, "Analyzing Educational Policies."
11. Kara Jackson et al., "Instructional Improvement and Instructional Management: District Leaders' Orientations Towards Improving Mathematics Teaching and Learning" (paper presented at the annual meeting of the *University Council for Educational Administration*, Washington, D.C., November 20–23, 2014).
12. Ibid.
13. Robert D. Benford and David A. Snow, "Framing Processes and Social Movements: An Overview and Assessment," *Annual Review of Sociology* 26 (2000): 611–639; Cynthia E. Coburn, "Framing the Problem of Reading Instruction: Using Frame Analysis to Uncover the Microprocesses of Policy Implementation," *American Educational Research Journal* 43, no. 3 (2006): 343–379.
14. Jackson et al., "Instructional Improvement and Instructional Management."
15. Our findings regarding district leaders articulating "form-level" instructional visions are similar to those described in James P. Spillane, "Cognition and Policy Implementation: District Policymakers and the Reform of Mathematics Education," *Cognition and Instruction* 18, no. 2 (2000): 141–179. In addition, as described in chapter 12, we found that most principals articulated a form-level vision of mathematics instruction.
16. See also James P. Spillane and Charles L. Thompson, "Reconstructing Conceptions of Local Capacity: The Local Education Agency's Capacity for Ambitious Instructional Reform," *Educational Evaluation and Policy Analysis* 19 (1997): 185–203.
17. Lea Hubbard, Hugh Mehan and Mary Kay Stein, *Reform as Learning* (New York: Routledge, 2008).
18. David K. Cohen and Carol A. Barnes, "Pedagogy and Policy," in *Teaching for Understanding: Challenges for Policy and Practice*, eds. David K. Cohen, Milbrey W. McLaughlin and Joan E. Talbert (San Francisco, CA: Jossey-Bass, 1993): 207–239. See also Cobb and Jackson, "Analyzing Educational Policies."
19. Mary Kay Stein and Suzanne Lane, "Instructional Tasks and the Development of Student Capacity to Think and Reason: An Analysis of the Relationship Between Teaching and Learning in a Reform Mathematics Project," *Educational Research and Evaluation* 2, no. 1 (1996): 50–80.
20. Brooks A. Rosenquist, Anne Garrison Wilhelm and Thomas M. Smith, "The IQA Observational Rubric and Teacher Value-Added: Looking for Nonlinearities to Inform Practice and Policy" (paper presented at the annual meeting of the *American Educational Research Association*, Philadelphia, Pennsylvania, April 3-7, 2014).
21. Richard Elmore argued for the importance of school and district leaders in formal positions of authority identifying relevant forms of expertise and recognizing and collaborating with colleagues who have already developed those capabilities. See Richard F. Elmore, "Leadership as the Practice of Improvement" (paper presented at the *OECD International Conference on Perspectives on Leadership for Systemic Improvement*, London, England, July 6, 2006).
22. Ample research has identified the need for senior central office leaders to "set

direction." See, for example, Philip Hallinger, "Leadership for Learning: Lessons From 40 Years of Empirical Research," *Journal of Educational Administration* 49, no. 2 (2011): 125–142, doi.org/10.1108/09578231111116699; Hanna Kurland, Hilla Peretz, and Rachel Hertz-Lazarowitz, "Leadership Style and Organizational Learning: The Mediate Effect of School Vision," *Journal of Educational Administration* 48, no. 1 (2010): 7–30, doi.org/10.1108/09578231011015395; James Sebastian and Elaine Allensworth, "The Influence of Principal Leadership on Classroom Instruction and Student Learning: A Study of Mediated Pathways to Learning," *Educational Administration Quarterly* 48, no. 4 (2012): 626–663; Wallace Foundation, *School Principal as Leader: Guiding Schools to Better Teaching and Learning* (New York, NY: Wallace Foundation, 2010).

23. For a discussion of the activity, see Charlotte Dunlap et al., "Schooling Leaders on the Common Core," *Phi Delta Kappan Common Core Writing Project* (2015), http://www.kappancommoncore.org/schooling-leaders-on-the-common-core/.
24. Ruth Chung Wei et al., *Professional Learning in the Learning Profession: A Status Report on Teacher Development in the United States and Abroad* (Dallas, TX: National Staff Development Council, 2009).
25. In proposing that district leaders take an educative stance to their work with school personnel, we follow Meredith. I. Honig, "District Central Office Leadership as Teaching."
26. For examples of other research that highlights the impact of central office leaders' visions and practices on schools, see H. Dickson Corbett and B.L. Wilson, "The Central Office Role in Instructional Improvement," *School Effectiveness and School Improvement* 3, no. 1 (1992); 45–68; Robert E. Floden et al., "Instructional Leadership at the District Level: A Closer Look at Autonomy and Control," *Educational Administration Quarterly* 24, no. 2 (1988): 96–124; Kenneth Leithwood, "Characteristics of School Districts That are Exceptionally Effective in Closing the Achievement Gap," *Leadership and Policy in Schools* 9, no. 3 (2010): 245–291; and Andrea K. Rorrer, Linda Skrla and James Joseph Scheurich, "Districts as Institutional Actors in Educational Reform," *Educational Administration Quarterly* 44, no. 3 (2008): 307–357. For an example of how central office leaders' vision matters especially for English learners, see Ana M. Elfers and Thomas Stritikus, "How School and District Leaders Support Classroom Teachers' Work with English Language Learners," *Educational Administration Quarterly* 50, no. 2 (2014): 305–344.
27. See also Patricia Burch and James P. Spillane, "How Subjects Matter in District Office Practice: Instructionally Relevant Policy in Urban School District Redesign," *Journal of Educational Change* 6, no. 1 (2005): 51–76; Meredith. I. Honig, "District Central Office Leadership as Teaching."
28. Honig, "District Central Office Leadership as Teaching."

CHAPTER 14

1. For additional insights about how researchers can support the work of school districts, Marco Muñoz and Angela Harris, two of our district leader partners, discuss the value of the MIST partnership to their district and consider the implications for researchers who want to influence practice in a white paper published on the MIST website. See Marco A. Muñoz and Angela M. Harris,

Research-Practice Partnerships in Action: A District Perspective, Retrieved from http://www.vanderbi.lt/mist.

CHAPTER 15

1. John D. Bransford et al., *How People Learn: Brain, Mind, Experience, and School* (Washington, D.C.: National Academy Press, 2000); Grant Wiggins and Jay McTighe, *Understanding By Design* (Washington, D.C.: Association for Curriculum and Supervision, 1998).
2. James P. Spillane, *Distributed Leadership* (San Francisco: Jossey Bass, 2005); James P. Spillane, Richard Halverson, and John B. Diamond, "Towards a Theory of Leadership Practice: Implications of a Distributed Perspective," *Educational Researcher* 30, no. 3 (2001): 23-30.
3. Richard F. Elmore, "Leadership as the Practice of Improvement," (paper presented at the OECD International Conference on Perspectives on Leadership for Systemic Improvement, London, England, July 6, 2006), 21.
4. Ibid.
5. Michael Fullan, *Choosing the Wrong Drivers for Whole System Reform* (Melbourne, Australia: Centre for Strategic Education, 2011).
6. See, for example, James E. Tarr et al., "The Impact of Middle-Grades Mathematics Curricula and the Classroom Learning Environment on Student Achievement, *Journal for Research in Mathematics Education* 39, no. 3 (2008): 247–280; Julie E. Riordan and Pendred E. Noyce, "The Impact of Two Standards-Based Mathematics Curricula on Student Achievement in Massachusetts, " *Journal for Research in Mathematics Education*, 32, no. 4 (2001): 368–398; Alan Schoenfeld, "Making Mathematics Work for All Children: Issues of Standards, Testing, and Equity," *Educational Researcher* 31, no. 1 (2002): 13–25.
7. This example is elaborated in Paul Cobb and Kara Jackson, "Analyzing Educational Policies: A Learning Design Perspective," *The Journal of the Learning Sciences* 21, no. 4 (2012): 487–521.
8. See Kara Jackson et al., "Investigating the Development of Mathematics Leaders' Capacity to Support Teachers' Learning on a Large Scale," *ZDM Mathematics Education* 47, no. 1 (2015): 93–104.
9. The design research methodology is especially appropriate for this and for other studies that involve clarifying learning goals and investigating how people can be supported to attain those goals.
10. Anthony S. Bryk and Louis Gomez, "Reinventing a Research and Development Capacity,"in *The Future of Educational Entrepreneurship: Possibilities for School Reform*, ed. Frederick M. Hess (Cambridge, MA: Harvard Education Press, 2008), 181–206; Anthony S. Bryk et al., *Learning to Improve* (Cambridge, MA: Harvard Education Press, 2015); David Yeager et al., *Practical Measurement* (Stanford, CA: Carnegie Foundation for the Advancement of Teaching, 2013).
11. For examples of research that provide evidence that teachers' instructional practices are profoundly influenced by the school and district settings in which they work, see Anthony S. Bryk et al., *Organizing Schools for Improvement: Lessons From Chicago* (Chicago: University of Chicago Press, 2010); Paul Cobb et al., "Situating Teachers' Instructional Practices in the Institutional Setting of the School and

School District," *Educational Researcher* 32, no. 6 (2003): 13–24; Cynthia E. Coburn, "Rethinking Scale: Moving Beyond Numbers to Deep and Lasting Change," *Educational Researcher* 32 no. 6 (2003): 3–12; Richard F. Elmore, *School Reform From the Inside Out* (Cambridge, MA: Harvard Education Press, 2004).

12. See, for example, Cynthia E. Coburn and William R. Penuel, "Research-Practice Partnerships in Education: Outcomes, Dynamics, and Open Questions," *Educational Researcher* 45, no. 1 (2016): 48–54, doi: 10.3102/0013189X16631750; National Research Council, *Strategic Education Research Partnership,* M. Suzanne Donovan, A.K. Wigdor, and Catherine E. Snow, editors. Division of Behavioral and Social Sciences and Education. (Washington, DC: The National Academies Press, 2003); Vivian Tseng, *Partnerships: Shifting the Dynamics Between Research and Practice* (New York: William T. Grant Foundation, 2012).

AFTERWORD

1. Data are first administration of the 8th grade Texas Exit math test. These include the Texas Assessment of Academic Skills, and the State of Texas Assessments of Academic Readiness. Data can be retrieved at: https://rptsvr1.tea.texas.gov/perfreport/aeis/ and https://rptsvr1.tea.texas.gov/perfreport/tapr/
2. Cynthia E. Coburn and William R. Penuel, "Research-Practice Partnerships in Education: Outcomes, Dynamics and Open Questions," *Educational Researcher* 45, no. 1 (2016): 48–54.

APPENDIX

1. Cynthia E. Coburn, William R. Penuel, and Kimberly Geil, *Research-Practice Partnerships at the District Level: A New Strategy for Leveraging Research for Educational Improvement* (Berkeley, CA and Boulder, CO: University of California and University of Colorado, 2013).
2. Hugh Burkhardt and Alan H. Schoenfeld, "Improving Educational Research: Toward a More Useful, More Influential, and Better-Funded Enterprise," *Educational Researcher* 32, no. 9 (2003): 3–14.
3. M. Suzanne Donovan, "Generating Improvement through Research and Development in Educational Systems," *Science* 340 (2013): 317–19.
4. William R. Penuel and Caitlin Farrell, "Research-Practice Partnerships and ESSA: A Learning Agenda for the Coming Decade," (in press) in *The Social Side of Reform,* ed. Esther Quintero-Corrall. (Cambridge, MA: Harvard Education Press).
5. Ron Haskins and Greg Margolis, *Show Me the Evidence: Obama's Fight for Rigor and Results in Social Policy* (Washington, DC: Brookings Institution, 2015).
6. Cynthia E. Coburn, Judith Toure, and Mika Yamashita, "Evidence, Interpretation, and Persuasion: Instructional Decision Making at the District Central Office," *Teachers College Record* 111, no. 4 (2009): 1115-61. Damien Contandriopoulos, et al., "Knowledge Exchange Processes in Organizations and Policy Arenas: A Narrative Systematic Review of the Literature," *The Milbank Quarterly* 88, no. 4 (2010): 444–83.
7. Coburn, *Research-Practice Partnerships at the District Level.*
8. Burkhardt and Schoenfeld, "Improving Educational Research". Suzanne M. Donovan, A. K. Wigdor, and Catherine E. Snow, *Strategic Education Research Partnership* (Washington, DC: National Research Council, 2003); Ann E. Kazak, et al., "A

Meta-Systems Approach to Evidence-Based Practice for Children and Adolescents," *American Psychologist* 65, no. 2 (2010): 85–97.

9. William R. Penuel, Cynthia E. Coburn, and Dan Gallagher, "Negotiating Problems of Practice in Research-Practice Partnerships Focused on Design," in *Design-Based Implementation Research: Theories, Methods, and Exemplars. National Society for the Study of Education Yearbook*, ed. Barry J. Fishman et al. (New York, NY: Teachers College Record, 2013), 237–55.
10. Dennis A. Gioia and Kumar Chittipeddi, "Sensemaking and Sensegiving in Strategic Change Initiation," *Strategic Management Journal* 12, no. 6 (1991): 433–48.
11. Coburn, *Research-Practice Partnerships at the District Level.*

ACKNOWLEDGMENTS

WE FIRST NEED TO THANK the district leaders, school leaders, instructional coaches, and teachers in our four partner districts. Without their time and commitment to this project and their desire to improve middle-grades mathematics teaching and learning, this project would not have been possible. Over the years, their openness, candor, and support allowed us to investigate the complex endeavor of instructional improvement. We would also like to thank the Institute for Learning and Victoria Bill for assistance in identifying and recruiting large urban districts that were aiming to support all students' attainment of rigorous learning goals in mathematics.

When we first started the collaboration that became MIST, we envisioned a relatively small project. We never imagined that the MIST Team would grow to include such a diverse group of staff, students, and faculty. For most MIST team members, this project was much more than a research study. As a research team, we worked and traveled together through thick and thin. We missed flights, lost luggage, and drove far too many miles in rental cars. We got lost, skipped meals, and spent numerous hours in hotels and airports together. We also got to know four cities, and visited many of their coffee shops and restaurants. Most importantly, we got to know and become colleagues with dedicated professionals who were working to improve educational opportunities for students in their cities.

We are indebted to those who joined us on this journey. Although unconventional, this is why it was both appropriate and essential to invite members of the team to contribute to the book, and to list *The MIST Team* as an author. This project was successful because of the hard work and expertise of many people over the years.

Melissa Boston: for her expertise and ongoing training on using the Instructional Quality Assessment rubrics and her collaboration on designing professional development for school leaders and mathematics coaches.

Ken Frank: for an ongoing research collaboration to investigate the relationship between teachers' advice networks and instructional improvement in mathematics.

Ilana Horn: for her contributions as a co-principal investigator for four years (2011–2015) and for an ongoing research collaboration to investigate the relationship between teacher collaborative meetings and instructional improvement in mathematics.

Post-doctoral research fellows: Mollie Appelgate, Dan Berebitsky, Christine Andrew-Larson, and Jessica Rigby. Their analyses and insights appear throughout this book.

Much of the heavy lifting on MIST was performed by doctoral students. A mixed methods, design-based project requires people who have a wide range of expertise—often developed as they participate in the research itself. Many thanks to Jason Brasel, I-Chien Chen, Glenn Colby, Brette Garner, Lynsey Gibbons, Sarah Green, Seth Hunter, Britnie Kane, Karin Katterfeld, Emily Kern, Nicholas Kochmanski, Adrian Larbi-Cherif, Charles Munter, John Murphy, Mahtab Nazemi, Brooks Rosenquist, Rebecca Schmidt, Charlotte Sharpe, Elizabeth Smith, Min Sun, Megan Webster, Anne Garrison Wilhelm, Jonee Wilson, Jana Visnovska, and Qing Zhao—without whom MIST would not have been possible.

Over the years, many undergraduate and master's students also contributed to the MIST project as research assistants. Several worked with us for multiple years and made a significant contribution, including Candice Clark, Denice Kelley, Emily Partin, Jordan Ridge, Shena Sanchez, and Jonathon Wong.

Because our districts were not in Tennessee, we had to hire video-data collectors to record classroom lessons each year. We are grateful to have worked with a group of people who were interested in contributing to an effort to improve instruction. They represented the project in a professional manner and helped to maintain relationships with school leaders and teachers.

Each summer we hired graduate students to code classroom video-recordings using the Instructional Quality Assessment. These coders attended a three-day training, and worked tirelessly to maintain reliability across the team to ensure that our data was high quality.

We are grateful to Caroline Chauncey for her ongoing guidance and support as our editor at *Harvard Educational Press.* Her astute eye and invariably constructive feedback pushed our thinking and enabled us to produce a book that captures our collective efforts over the past ten years.

Finally, we would like to thank our families for supporting us in innumerable ways over the course of this project.

The empirical work reported in this book has been supported by the National Science Foundation (NSF) under grants DRL-0830029; ESI-0554535; and DRL-1119122. A portion of funding for several graduate students was supported by the Institute of Education Sciences (IES) predoctoral research training program, grant number R305B080025. Two postdoctoral research fellows were supported by the Institute for Education Sciences (IES) of the U.S. Department of Education postdoctoral training, grant number R305B080008. The National Academy of Education/Spencer Postdoctoral Fellowship Program supported Kara Jackson's research related to views of students' current mathematical capabilities. Any opinions, findings, and conclusions or recommendations expressed in this book are those of the authors and do not necessarily reflect the views of our funders.

Paul Cobb
Kara Jackson
Erin Henrick
Thomas M. Smith

ABOUT THE AUTHORS

PAUL COBB is a research professor at Vanderbilt University. He is the principal investigator of the MIST study. He received his doctorate in mathematics from The University of Georgia. He has taught courses in elementary mathematics teaching, design research methodology, and the school and district setting of mathematics teaching and learning. His research focuses on improving the quality of mathematics teaching and student learning on a large scale, and on issues of equity in students' access to significant mathematical ideas.

KARA JACKSON is an associate professor of mathematics education at the University of Washington, Seattle. She received her doctorate in Education, Culture, and Society with an emphasis in mathematics education at the University of Pennsylvania Graduate School of Education. Dr. Jackson is a co-principal investigator of MIST. Her research focuses on specifying forms of practice that support all learners to participate in rigorous mathematics and how to organize educational contexts to support teachers to develop such practices.

ERIN HENRICK is a Senior Research Associate in the Department of Teaching and Learning at Vanderbilt University. She is the project manager and is a co-principal investigator on the MIST project, and received her doctorate in Education, Leadership, and Policy from Vanderbilt University. Dr. Henrick's research interests include district-level design research, researcher practitioner partnerships that support improvements in the quality of teaching, and school and district supports to foster instructional improvement.

DR. THOMAS M. SMITH is the Dean of the Graduate School of Education at the University of California, Riverside, and is a co-principal investigator of the MIST project. He earned his doctorate in education theory and policy from the Pennsylvania State University. Previously he was associate professor and Director of Graduate Studies in the Department of Leadership, Policy, and Organizations at Vanderbilt University. His research focuses on school and district instructional improvement in mathematics, and on research practitioner partnerships.

ABOUT THE CONTRIBUTORS

ANNA RUTH ALLEN is a research associate in the School of Education at the University of Colorado Boulder. Her research has focused on learning, literacy, youth development, and understanding and supporting relationships between research and practice. She has conducted qualitative studies of learning in educational contexts in and out of schools, as well as studies of research-practice partnerships in urban districts. She currently contributes to the IES-funded National Center for Research in Policy and Practice and the NSF-funded Research+Practice Collaboratory by studying the role of research evidence in educational decision making, and supporting collaborative partnerships between educators and researchers.

MOLLIE APPELGATE is an assistant professor of mathematics education at Iowa State University. Her research focuses on pre-service and in-service teacher STEM learning. She earned her doctorate in education from the University of California, Los Angeles. While working on the MIST project as a post-doctoral fellow, Dr. Appelgate investigated teachers' curriculum use in the classroom, in professional development, and during teacher collaborative time. She is currently investigating how early elementary teachers use mathematically focused text messages with linguistically diverse parents to better connect home and school mathematics and is also contributing to the development of STEM curriculum for elementary and middle school students that focuses on the engineering design process.

I-CHIEN CHEN is a doctoral candidate at Michigan State University and was a graduate research assistant on the MIST project for four years. She has investigated how the frequency and the quality of teacher collaborative meetings influence the formation of teacher advice networks, and whether accountability pressures stimulate teachers' collaboration through advice seeking. She has also studied how friendship network and family-school connections are associated with mathematics achievement and STEM majors among high school students using national representative data from NCES.

KENNETH FRANK is an MSU Foundation professor of Sociometrics, professor in Counseling, Educational Psychology and Special Education; Fisheries and Wildlife, and has an adjunct appointment (by courtesy) in Sociology at Michigan State University. He earned his doctorate in in measurement, evaluation and statistical analysis from the School of Education at the University of Chicago. His interests include the study of schools as organizations, social structures of students, teachers, and school decision making, and social capital. These areas are linked to several methodological interests: social network analysis, causal inference and multi-level models. Dr. Frank's current projects include studies of how schools respond to increases in core curricular requirements, cognitive linkages among aspects of knowledge, and how adolescents respond to their social contexts in schools.

BRETTE GARNER is a doctoral student in math education at Vanderbilt University. She has been a graduate research assistant on the MIST project for three years. Before coming to graduate school, she was a middle school math teacher in Texas and worked on developing assessments for the Common Core State Standards. Her research focuses on teacher learning in workgroup settings, with a particular interest in teachers' use of assessment data.

LYNSEY GIBBONS is an assistant professor of mathematics education at Boston University and was a graduate research assistant on the MIST project for five years. Her research aims to clarify how the contexts in which teachers work can be organized to support their ongoing learning, and how instructional leaders' and professional educators' practices can support teachers' development of high-quality mathematics instruction.

ILANA SEIDEL Horn is Professor of Mathematics Education at Vanderbilt University. After teaching high school in California, she earned her doctorate at the University of California, Berkeley. Dr. Horn is a co-PI on the MIST project. Her research examines secondary mathematics teachers' learning in the contexts of their workplace, yielding images of teachers' learning and practice that account for the institutional setting of schools and the pressures of policy. The goal of her research is to improve supports for teachers' professional learning as a strategy for improving students' experience in math class, particularly in urban schools.

BRITNIE DELINGER KANE is an Assistant Professor at the Citadel in Charleston. She earned her doctorate from Vanderbilt University and was a graduate research assistant on the MIST project for four years. Dr. Kane's research focuses on teachers' learning about and development of equitable and rigorous instructional practices. She has studied both pre-service and in-service teachers' learning, both in preservice teacher education programs and in in-service teachers' professional learning communities. Currently, Dr. Kane is studying how instructional coaches facilitate teachers' learning and how school and district contexts influence coaches' work with teachers.

ADRIAN LARBI-CHERIF is a post-doctoral education researcher at George Washington University in the Department of Education Leadership. He earned his doctorate from Vanderbilt University and was a graduate research assistant on the MIST project for five years. His research investigates instructional leadership and improving instruction and student learning at the scale of large districts. His dissertation examined what school leaders need to know and be able to do in order to provide math teachers with opportunities to improve their instruction.

NICHOLAS KOCHMANSKI is a doctoral student in mathematics education at Vanderbilt University. He has been a graduate assistant on the MIST project for three years. Prior to his doctoral studies, he worked as a middle school mathematics teacher, supported teachers as an instructional coach, and earned his Master of Public Policy degree from Vanderbilt University. His research interests include instructional coaching, supports for instructional coach learning, and the design of tools to support improvements in mathematics instruction at scale.

CHARLES MUNTER is an assistant professor of mathematics education at the University of Missouri and was a graduate research assistant on the MIST project for six years. His research focuses on defining high-quality mathematics instruction and understanding and supporting teachers' learning in particular school and district settings. A recent interest has been formative intervention research aimed at achieving racial equity in secondary mathematics in urban school districts.

MAHTAB NAZEMI is an assistant professor at Thompson Rivers University in Kamloops, BC, Canada. She earned her doctorate in curriculum and instruction from the University of Washington, Seattle. She has been involved with the MIST project since completing her Master's at McGill University in 2011. Mahtab started her teaching career as a high school Mathematics and French Immersion teacher. As well as consulting and coaching work, she has taught K-12 teacher education and methods courses at McGill University, Antioch University, and the University of Washington. Her research employs critical race and anti-racist feminist theories in order to understand neoliberalism and institutional racism in K-16 educational institutions and settings.

WILLIAM R. PENUEL is a professor of learning sciences and human development in the School of Education at the University of Colorado Boulder. He received his doctorate in developmental psychology from Clark University. His current research examines conditions needed to implement rigorous, responsive, and equitable teaching practices in STEM education. As Principal Investigator of the National Center for Research in Policy and Practice (http://www.ncrpp.org), Penuel leads a team studying how district and school leaders use research.

JESSICA G. RIGBY is an assistant professor in Education Policy, Organizations, and Leadership in P12 Systems at the University of Washington and was a post-doctoral fellow on the MIST project for two years. After teaching middle and high school in California, she earned her doctorate at the University of California, Berkeley. She draws on organizational sociology to understand the role of school and district leaders in implementing policy, improving teacher practice, enhancing equity in student learning opportunities, and influencing teachers' informal advice networks. Jessica's current work examines the organizational structures and learning needs that support school and district leaders' work for school improvement, specifically in mathematics.

CHARLOTTE SHARPE is an assistant professor of mathematics education at Syracuse University. She earned her doctorate in mathematics education from Peabody College at Vanderbilt University and was a graduate assistant on the MIST project for five years. Her research interests include teacher learning, professional development, and high-leverage means of supporting teachers' development of ambitious instructional practices. Her work on the MIST project included designing professional development of mathematics coaches, investigating the role of school leaders in supporting teachers' instructional improvement, and developing a measure of teachers' views of their students' mathematical capabilities.

MICHAEL SORUM served the Fort Worth Independent School District from 2005 until retiring in 2016, first as Chief Academic Officer and for the last five years as Deputy Superintendent responsible for academics, school leadership, special programs, and student support services. Prior to Fort Worth, Sorum taught for twelve years in San Antonio, Texas and served as the Director of Accountability and Chief Academic Officer in the Providence, Rhode Island School Department. In addition to the MIST project, Sorum also led the district components of long-term Research Practice Partnerships with Vanderbilt and the EDC on high school scale up of effective practices and Harvard University on Internal Coherence at the elementary school level. Sorum received his bachelor's in romance languages at Portland State University, his Master's in Urban Superintendency at Harvard, and his doctorate in Education Leadership at Texas Christian.

MEGAN WEBSTER is an educational consultant. She teaches pre-service teachers at McGill, trains professional development (PD) leaders at the Ministry of Education of Quebec, and consults on projects that aim to improve instruction at scale. She received her doctorate from McGill University and was a graduate assistant on the MIST project for four years. Dr. Webster's research focuses on how to support the learning of professional development leaders and facilitators, and on the impact of professional development for principals on teachers' opportunities to learn.

ANNIE WILHELM is an assistant professor of mathematics education in the Department of Teaching and Learning at Southern Methodist University. Dr. Wilhelm earned her doctorate at Vanderbilt University, where she was an IES Pre-Doctoral Fellow. She was a graduate assistant on the MIST project for five years. Dr. Wilhelm's scholarly interests focus on mathematics teachers' knowledge and practice, supports for teacher development, and measuring complex practices.

JONEE WILSON is an assistant professor of Mathematics Education at North Carolina State University. After teaching high school in Baltimore, she earned her doctorate in 2015 at Vanderbilt University. Her research examines supports for historically marginalized students. She was a graduate assistant on the MIST project for five years, and worked for one additional year as a post-doctoral fellow. Her research examines supports for historically marginalized students. Specifically, she is interested in identifying forms of instructional practice that support students who are English Language Learners and low-performing African American students to participate and succeed in rigorous/inquiry oriented mathematics. She is also interested in specifying how to design and lead professional development that supports teachers in developing, implementing, and sustaining such instructional practices.

INDEX